T0361264

Machine Learning for Business Analytics

Machine learning is an integral tool in a business analyst's arsenal because the rate at which data is being generated from different sources is increasing and working on complex unstructured data is becoming inevitable. Data collection, data cleaning, and data mining are rapidly becoming more difficult to analyze than just importing information from a primary or secondary source. The machine learning model plays a crucial role in predicting the future performance and results of a company. In real time, data collection and data wrangling are the important steps in deploying the models. Analytics is a tool for visualizing and steering data and statistics. Business analysts can work with different data sets—choosing an appropriate machine learning model results in correct analyzing, forecasting the future, and making informed decisions. The global machine learning market was valued at $1.58 billion in 2017 and is expected to reach $20.83 billion in 2024—growing at a CAGR of 44.06% between 2017 and 2024. The authors have compiled important knowledge on machine learning real-time applications in business analytics. This book enables readers to get broad knowledge in the field of machine learning models and to carry out their future research work. The future trends of machine learning for business analytics are explained with real case studies. Essentially, this book acts as a guide to all business analysts. The authors blend the basics of data analytics and machine learning and extend its application to business analytics. This book acts as a superb introduction and covers the applications and implications of machine learning. The authors provide first-hand experience of the applications of machine learning for business analytics in the section on real-time analysis. Case studies put the theory into practice so that you may receive hands-on experience with machine learning and data analytics. This book is a valuable source for practitioners, industrialists, technologists, and researchers.

Machine Learning for Business Analytics

Real-Time Data Analysis for Decision-Making

Edited by
Hemachandran K., Sayantan Khanra,
Raul V. Rodriguez & Juan R. Jaramillo

A PRODUCTIVITY PRESS BOOK

First published 2023
by Routledge
605 Third Avenue, New York, NY 10158

and by Routledge
4 Park Square, Milton Park, Abingdon, Oxon, OX14 4RN

Routledge is an imprint of the Taylor & Francis Group, an informa business

© 2023 selection and editorial matter, Hemachandran K., Sayantan Khanra, Raul V. Rodriguez & Juan R. Jaramillo; individual chapters, the contributors

The right of Hemachandran K., Sayantan Khanra, Raul V. Rodriguez & Juan R. Jaramillo to be identified as the authors of the editorial material, and of the authors for their individual chapters, has been asserted in accordance with sections 77 and 78 of the Copyright, Designs and Patents Act 1988.

All rights reserved. No part of this book may be reprinted or reproduced or utilised in any form or by any electronic, mechanical, or other means, now known or hereafter invented, including photocopying and recording, or in any information storage or retrieval system, without permission in writing from the publishers.

Trademark notice: Product or corporate names may be trademarks or registered trademarks, and are used only for identification and explanation without intent to infringe.

Library of Congress Cataloging-in-Publication Data
A catalog record for this title has been requested

ISBN: 978-1-032-07281-4 (hbk)
ISBN: 978-1-032-07277-7 (pbk)
ISBN: 978-1-003-20631-6 (ebk)

DOI: 10.4324/9781003206316

Typeset in Garamond
by Apex CoVantage, LLC

Contents

Preface

Machine learning is increasingly recognized for playing a crucial role in predicting business performances for firms across the world. On the one hand, data collection and data wrangling are the important steps in training machine learning models. On the other hand, business analytics includes various tools for visualizing and steering data to deliver insights in management decisions. An expert business analyst may work with different data sets, choose appropriate machine learning models, and deliver valuable insights to improve business performances.

A report published by Market Research Future predicts that the global machine learning market will grow at over 44% compound annual growth rate till 2024 to reach $20.83 billion, up from $1.58 billion in 2017. Needless to say, the adoption of machine learning in business enterprises would create enormous opportunities for students and professionals with the required skill set. This book aims to help machine learning enthusiasts better prepare for the opportunity.

Overall, this book captures applications of machine learning in different contexts, such as agriculture, B2B marketing, banking, customer experience, digital marketing, healthcare, hospitality, insurance, and sustainable development. The book is comprised of 13 chapters.

We take this opportunity to express our sincere gratitude to the authors of the 13 chapters in this book. The authors include academicians from renowned institutions and industry experts with relevant experience. We believe that this book will be of significant value to different interest groups that wish to explore the applications of machine learning. Finally, we wholeheartedly thank our editorial team for extending persistent support to publish this book.

Editors' Biographies

Hemachandran K.

Dr. Hemachandran K. is a Professor of Artificial Intelligence at the School of Business, Woxsen University, Hyderabad, Telangana, India. He is a passionate teacher with 14 years of teaching experience and 5 years of research experience. He is a strong educational professional with a scientific bent of mind, highly skilled in AI and ML. After receiving his PhD in embedded systems from Dr. MGR Educational & Research Institute, India, he conducted interdisciplinary research in AI. He is an open-ended positive person who has a stupendous peer-reviewed publication record with more than 20 journals and international conference publications. He served as an effective resource person at various national and international scientific conferences. He has rich research experience in mentoring UG and PG student projects. He owns two patents and has life membership in esteemed professional organizations. He was a pioneer to establish the Single Board Computer Lab at Ashoka Institutions, Hyderabad, India. His self-paced learning schedule and quest to upgrade and update learning skills resulted in receipt of 15 online certificate degrees conferred by COURSERA and other online platforms. He is also an editorial board member for numerous reputed SCOPUS/SCI journals.

Sayantan Khanra

Dr. Sayantan Khanra serves as Assistant Professor at the School of Business Management, Narsee Monjee Institute of Management Studies, Mumbai, India. He holds a PhD in strategic management from the Indian Institute of Management Rohtak. His research interests relate to the domains of the digital economy, management of technology, and sustainable development. Previously, he worked at the Turku School of Economics, Finland; the National Taiwan University of Science and Technology, Taipei; and Woxsen University, India. His research studies are published in *Business Strategy and the Environment*, *Enterprise Information Systems* and the *Journal of Business Research*, among other quality academic journals.

Raul V. Rodriguez

Dr. Raul V. Rodriguez is Pro-Vice-Chancellor at Woxsen University and Dean of the School of Business at Woxsen University. He holds a PhD in artificial intelligence and robotics process automation applications in Human Resources. He is former Co-CEO of Irians Research Institute, a research facility specialized in neuromarketing, AI, ML, market research, behavioral science, social research, and behavioral engineering. His areas of expertise and interest are machine learning, deep learning, natural language processing, computer vision, robotic process automation, multi-agent systems, knowledge engineering, and quantum artificial intelligence. He is proficient in Prolog, Java, C++, Python, R/RStudio, Julia, Swift, Scala, MySQL, and Spark, among others. He is a registered expert in artificial intelligence, intelligent systems, and multi-agent systems at the European Commission, a nominee for the Forbes 30 under 30 Europe 2020 list, and an awardee in the Europe India 40 under 40 leaders. Additionally, he is a member of the GRLI Deans and Directors cohort. He has co-authored two reference books: *New Age Leadership: A Critical Insight* and *Retail Store'e* and has more than 70 publications to his credit. He is a weekly contributing writer to various magazines in the field of analytics and emerging technologies. He is also a journal reviewer and associate editor in various publications such as IEEE.

Juan R. Jaramillo

Juan R. Jaramillo is Associate Professor and Academic Director of the Master in Business Analytics in the Robert B. Willumstad School of Business at Adelphi University. He has a PhD and an MS in industrial engineering from West Virginia University, and he earned a BS in civil engineering and a BS in geological engineering from Escuela de Ingeniería de Antioquia in Medellín Colombia. He is also a Lean Six Sigma Master Black Belt. At Adelphi, Juan led the design of the Master of Science in Business Analytics (MSBA). He previously worked at Farmingdale State College in New York and Albany State University in Georgia. In these institutions, he designed an undergraduate program in business analytics and two supply chain programs. Early in his career Juan worked for Colceramica S. A., one of the largest ceramic producers in the Americas, in the areas of R&D and manufacturing. He has been involved in the Informs Innovative Applications in Analytics Award (IAAA) since its inception in 2012 and has served as a judge, chair (2017–2019), and cochair (2020–2022). He received the inaugural Michael Gorman Award for his contribution to the Analytics Society of INFORMS in 2020. His areas of expertise are analytics, artificial intelligence, and operations. He has multiple publications in these areas, and he has been the keynote speaker and panelist at conferences in the United States, Asia, and Latin America.

Contributors

M. Anand
Department of ECE
Dr. MGR Educational & Research Institute University
Chennai, India

Meghraj Arli
School of Business
Woxsen University
Hyderabad, India

Darsana S. Babu
Department of ECE
Baselios Mathews II College of Engineering
Sasthamcotta, Kollam, Kerala, India

Juvairia Begum
Department of CSE
LORDS Institute of Engineering & Technology
Hyderabad, India

Debdutta Choudhury
School of Business
Woxsen University
Hyderabad, India

S. Deepajothi
Department of CSE
Nagarjuna College of Engineering and Technology
Bangalore, India

Viplav Dhandhukia
Business Analyst
Simply Fresh Pvt. Ltd.
Hyderabad, India

Bijay Kumar G.
School of Business
Woxsen University
Hyderabad, India

Sanjeev Ganguly
School of Business
Woxsen University
Hyderabad, India

Preetha Mary George
Dr. MGR Educational Research Institute University
Chennai, India

C. Guzmán-Velásquez
Universidad EAFIT
Medellin, Colombia

L. K. Indumathi
Department of CSE
LORDS Institute of Engineering & Technology
Hyderabad, India

Syed Hasan Jafar
School of Business
Woxsen University
Hyderabad, India

Muneza Kagzi
Assistant Professor, Strategy & Sustainability
T. A. Pai Management Institute
Manipal, Karnataka, India

A. Kannan
Dean, Energy Conservation and Electrical Management
Dr. MGR Educational & Research Institute University
Chennai, India

Chinnapani Kiran Kumar
School of Business
Woxsen University
Hyderabad, India

Korupalli V. Rajesh Kumar
School of Business
Woxsen University
Hyderabad, India

J. G. Lalinde-Pulido
Universidad EAFIT
Medellin, Colombia

Megha Mankal
School of Business
Woxsen University
Hyderabad, India

Praveen Kumar Munari
School of Business
Woxsen University
Hyderabad, India

Gaurav Nagpal
Birla Institute of Technology and Science
Pilani, India

B. Justus Rabi
Principal
Christian College of Engineering and Technology
Tamilnadu, India

H. Raghupathi
Department of ECE
Visvesvaraya College of Engineering and Technology
Hyderabad, India

Abdul Rais
Department of CSE
LORDS Institute of Engineering & Technology
Hyderabad, India

Princy Sera Rajan
Department of ECE
Baselios Mathews II College of Engineering
Sasthamcotta, Kollam, Kerala, India

Namita Ruparel
Department of HR
IBS Hyderabad
IFHE University
Hyderabad, India

Sameena M. H.
Department of AEI
Baselios Mathews II College of Engineering
Sasthamcotta, Kollam, Kerala, India

Thakur Santosh
School of Technology
Woxsen University
Hyderabad, India

Meganathan Kumar Satheesh
Panasonic Life Solutions India Pvt Ltd
Chennai, India

Himanshu Seth
Indian Institute of Management
Rohtak, India

Cynthia Jabbour Sfeir
Office of Finance Notre Dame University-Louaize NDU
Zouk Mosbeh, Lebanon

Laxmi Shaw
Assistant Professor
Department of ECE
Chaitanya Bharati Institute of Technology
Hyderabad, India

Dr. M. A. Sikandar
School of Commerce and Business Management
Maulana Azad National Urdu University
Hyderabad, India

Kshitiz Sinha
Analyst
IQVIA
BAaaS Consultant, IQVIA,
Bengaluru, India

Harshitha Sirineni
School of Business
Woxsen University
Hyderabad, India

Venkat Reddy Yasa
School of Business
Woxsen University
Hyderabad, India

Chapter 1

Introduction to Machine Learning for Data Analytics

L. K. Indumathi, Abdul Rais, and Juvairia Begum

Contents

DOI: 10.4324/9781003206316-1

1.1 Introduction

In the 1960s, computer science became an academic discipline. Many basic computer science subjects like computer architecture, operating systems, and computer networks, underpins them and were all highlighted. The study of algorithms was added as an important component of theory in the 1970s. The goal was to make computers more useful. Today, a fundamental shift is taking place, with the emphasis shifting to a wide range of applications. There are a variety of explanations behind this shift. Computing and communications have become increasingly intertwined. In the natural sciences, commerce, and other sectors, the increased ability to monitor, acquire, and store data necessitates a shift in our understanding of data and how to handle it in the modern era. The rise of the internet and social media as important components of daily life brings theoretical opportunities as well as challenges.

Whereas traditional areas of computer science will continue to be important, future researchers will be more concerned with using computers to understand and extract usable data from multiple data generated by applications, rather than just how to make computers helpful for specific well-defined problems.

Digital data is frequently presented with a large number of components in domains as varied as cognition, retrieval, and machine learning (ML). The topic model is more than just a way to keep track of multiple fields in a record.

One of the most surprising developments in computer science in the last two decades is that some domain-independent methods have proven to be quite effective in solving problems from a variety of fields. A good example is ML.

Analysts can derive insights from data through statistical analysis. Data is analyzed using both statistics and ML approaches. Big Data is employed in the development of statistical models that show data trends. These models can then be used to create predictions and inform decision-making using new data. This procedure requires statistical programming languages like R or Python (with pandas). Advanced analysis is also possible because to open source libraries and packages like Tensor Flow.

The process of analyzing, cleaning, manipulating, and modeling data with the objective of identifying usable information, informing conclusions, and assisting decision-making is known as data analysis. Data analysis has several dimensions and approaches, including a wide range of techniques under various titles and being applied in a variety of business, science, and social science sectors. Data analysis is important in today's business environment since it helps businesses make more scientific decisions and run more efficiently.

Chapter 1 gives an overall view of data analysis, data processing, data cleaning, data visualizing, requirement of ML, probability of ML, and basic algorithm of ML.

1.2 Basics of Data

1.2.1 What Is Data?

Data in computing knowledge has been converted into a format that is easy to transfer or process. Data is information translated into binary digital form, as it relates to today's computers and transmission media. It is allowed to use data as either a solitary or plural subject. The term "raw data" refers to data in its most basic digital version.

The terms "data processing" and "electronic data processing," although for a time were used interchangeably to refer to the entire range of what was then recognized as digital technologies, indicated that data analysis is important in computer-supported collaborative. In the history of corporate computing, specialization has occurred, and a distinct data profession has emerged in line only with the development of enterprise data handling.

Computers represent data like video, images, audio, and text as binary values, which are made up of simply two numbers: 1 and 0. A bit is the simplest data unit, with a single value. Eight binary digits make up a byte. Megabytes and gigabytes are capacity and storage units.

As the amount of data collected and stored expands, so do the units of data measurement. For example, the phrase "brontobyte" refers to data storage equal to 10 to the 27th power of bytes.

Information will be stored in file types, identical to just how mainframe systems employ ISAM and VSAM. Some other data format for data storage, transmission, and analysis is comma-separated values. Further specialization occurred as a database, a database management system, and then relational database technologies appeared to organize information.

Over the last decade, the rise of the internet and smartphones has resulted in a boom in digital data production. Text, audio, and video data, as well as register and online activity records, are now included in the data. Unstructured data makes up a large portion of this.

Data of the petabyte or larger range has been referred to as "Big Data." The 3Vs—volume, variety, and velocity—describe large data in a simplified way. Big Data–driven business models have evolved as web-based e-commerce has increased in popularity, recognizing data as an important commodity for its own sake.

Outside of its use in data-processing computing applications, data has a value. In electrical component connectivity and network communication, the term "data" is often distinguished from "control information," "control bits," and related expressions to describe the core substance of a transmission unit. Furthermore, in science, the term "data" refers to a collection of facts. This can be seen in finance, marketing, demographics, and healthcare, to name a few.

1.2.2 What Is Data Analysis?

Working with data to extract relevant information that can subsequently be utilized to make informed decisions is known as data analysis.

This concept is the foundation of data analysis. We are better able to make decisions when we can extract meaning from facts. And we live in an era where we have access to more data than ever before.

1.2.3 Why Is Data Analysis Required?

In the business world, data analysis is critical for understanding challenges and exploring data in meaningful ways. Data is nothing more than numbers and facts. Data analysis is the process of organizing, interpreting, structuring, and presenting data into valuable information.

Everyone realizes that the goal of data analysis is to help you make data-driven business choices, otherwise why would you let it take so long that the results are obsolete by the time you get them? Web data integration automates all processes of web data analysis, allowing you to gain insights from data as soon as it is collected. You can use real-time data insights instead of obsolete insights as a foundation for your company decisions.

1.2.3.1 Types of Data Analysis

Data can be utilized in a variety of ways to answer questions and assist choices. These types of analyses can be categorized into four groups that are regularly employed in the field. We'll go through each of these data analysis techniques, as well as an example of how they could be used in the real world.

1.2.3.1.1 Descriptive Analysis

Big Data and data science have become popular terms in recent years. They tend to be well-researched, which necessitates careful processing and analysis of the data. Descriptive analysis is one of the approaches used to analyze this data. What transpired is revealed through descriptive analysis. This sort of analysis uses statistics to describe or summarize quantitative data. Statistical analysis, for example, might reveal the distribution of sales among a group of students as well as the average marks per student. Explanation of "What Happened" is referred as descriptive analysis.

1.2.3.1.2 Diagnostic Analysis

The "what" is determined by descriptive analysis, whereas the "why" is determined by diagnostic analysis. Now, let us imagine a descriptive analysis reveals that a hospital is experiencing an extraordinary influx of patients. If you go deeper into the data, you might find that many of these individuals have the same virus symptoms. This diagnostic study can help you figure out if the inflow of patients was caused by an infectious pathogen—the "why." Explanation of "Why It Happened" is referred as diagnostic analysis.

1.2.3.1.3 Predictive Analysis

As of now, we've examined the methods of analysis that look at the past and draw conclusions. Predictive analytics makes predictions about the future based on data. You might notice that a particular product has had its strongest sales during the months of September and October each year, leading you to predict a similar high point during the future year using predictive analysis. Explanation of "Future Status (What May Happen?)" is referred as predictive analysis.

1.2.3.1.4 Prescriptive Analysis

Prescriptive analysis combines the findings of the preceding three forms of analysis to make ideas for how a corporation should proceed. Applying our analogy, this form of analysis might recommend a business strategy to capitalize on the accomplishment of the high-sale months while also identifying fresh growth prospects during the weaker months. Explanation of "What Is the Reaction" is referred as diagnostic analysis. It will support the decision-making mechanism.

1.2.4 Base of Data Mining

Data mining is a systematic analytical process for extracting information from huge amounts of original data. It is a discipline of computer engineering and analytics that uses sophisticated approaches to find hidden patterns in massive amounts of data. The automation of this type of analysis, which uses ML and database technologies to speed up the analytical process and locate knowledge which is more relevant and engaging, is one of its distinguishing features. Facts mining, contrary to its term, is not usually about extracting pure data from a mountain of data but rather about detecting relevant regularities that arise again from data set.

The term "data mining" refers to the process of extracting information from large amounts of data. To put it another way, data mining is the process of extracting knowledge from data. The information or knowledge obtained in this manner can be applied to any of the following purposes like market analysis, web surfing, control of production, and science exploration.

1.2.4.1 Data Processing

Although most people are familiar with the term "word processing," computers were designed for "data processing"—the organizing and manipulation of enormous volumes of quantitative data, or "number crunching" in computer jargon. Calculation of satellite orbits, weather forecasting, statistical analysis, and, in a more practical sense, business applications such as accounting, payroll, and billing are all instances of data processing.

1.2.4.1.1 Types of Data Processing

Depending on the purpose of the data, many types of data processing procedures are available. The five major methods of data processing are discussed in this book.

(1) Large-Scale Data Processing or Commercial Data Processing Large-scale data processing is a method of using relational databases in a commercial setting, which involves batch processing. It entails feeding the system a vast amount of data and producing a significant volume of output with fewer processing operations. It essentially integrates commerce and computers in order to make it helpful for a company. Because the data processed by this system is usually standardized, there is a significantly smaller risk of errors.

Many manual tasks are automated using computers to make them easier and more error-free. In the business world, computers are employed to turn raw data into information that is beneficial to the company. Accounting software is a good example of a data processing application. The field of information systems (IS) studies subjects like organizational computer systems.

(2) Research or Scientific Data Processing Scientific or research data processing, unlike commercial data processing, makes extensive use of computing operations while requiring fewer inputs and outputs. Arithmetic and comparison operations are among the computing operations. Any chance of errors is unacceptable in this form of processing since it would lead to erroneous decisions. As a result, the process of validating, categorizing, and standardizing the data is carried out with great care, and a variety of scientific procedures are employed to ensure that no incorrect associations or conclusions are made.

(3) Group Processing Group processing is a sort of data processing that involves processing multiple cases at the same time. It is most commonly utilized when the data is homogeneous and in big amounts, and it is collected and analyzed in batches. Ongoing, sequential, or chronological processing of an activity is referred to as batch processing. Simultaneous group processing occurs when all of the cases are processed at the same time by the same resource. Sequential batch processing occurs when various cases are processed by the same resource either simultaneously or sequentially.

(4) Electronic Procession The answer to each point and click is computed long before the user even opens the application in most electronic analytical processing systems. In truth, many online processing systems conduct the computation inefficiently, but because the processing is done ahead of time, the end-user is unaware of the issue. When data must be processed on a continuous basis and is fed into the system automatically, this sort of processing is utilized.

(5) Real-Time Processing The existing data management system often limits the capacity of processing data on an as-needed basis because it is always based on periodic batch updates, resulting in a time lag of many hours between an event occurring and it being recorded or updated. This necessitated the development of a system that could record, update, and process data on an as-needed basis, that is, in real-time, decreasing the time delay among event and execution to nearly zero. Huge amounts of data are being pumped into the systems of businesses; therefore, storing and analyzing it in real time would be a game changer.

1.2.4.1.2 Data Pre-processing

Pre-processing data is a data mining technique for transforming raw data into a usable and productive format. Data pre-processing entails converting raw data into well-formed data sets in order to use data mining methods. Raw data is frequently incomplete and formatted inconsistently. The success of every project involving data analytics is directly proportional to the quality of data preparation.

1.2.4.2 Data Cleaning

Data cleaning, also known as data scrubbing, is one of the most important features required to create an entity that contains the art of excellent decision-making. There's no denying that an analysis can only be good if the data it is based on is of high quality. The act of producing data for specification by removing or reshaping data that is missing, wrong, inappropriate, unsuitable, or redundant is known as data cleaning.

1.2.4.2.1 What Is Data Cleaning?

The practice of updating or removing incorrect, duplicate, corrupted, or incomplete data from a database is known as data cleaning. If data is wrong, even if it appears to be correct, algorithms and outputs are unreliable. The data cleaning process isn't only about eliminating data to make room for new data; it's also about figuring out how to maximize the validity of a data collection without erasing it.

Data cleaning entails correcting grammatical and spelling errors, repairing faults like omitting codes and empty fields, finding duplicate data points, and standardizing data sets, among other things. It is seen as a basic aspect of the data science fundamentals and plays a significant role in creating trustworthy results and in the analysis procedure. Data cleaning services are designed to create uniform and standardized data sets that give data analytical devices and predictive analytics good accessibility to and perception of precise figures for every challenge.

1.2.4.2.2 Comparison of Data Transformation and Data Cleaning

Data warehouses aid in data analysis, report creation, data visualization, and business decision-making. In data warehousing, two strategies are used: data transformation and data cleaning. Data cleaning refers to the removal of incoherent data from a database in order to improve data homogeneity, whereas data transformation refers to the translation of data from one structure to another in order to facilitate processing.

1.2.4.2.3 Method to Clean Data

A data cleaning method will change most parts of an entity's general data cleaning program, but it is only one part of a long-term data cleaning solution. The following are the steps for data cleaning

Step 1: The first step is to determine which types of data fields are required for the project.

Step 2: The information in the data fields that have been short-listed is gathered, categorized, and organized.

Step 3: Duplicate figures are identified and removed, and mistakes are corrected.

Step 4: Tools for data purification is to avoid information gaps, look for and fill in the missing values in the data collection.

Step 5: The data operation should be standardized based on repeated tests and methodologies that have shown to create high-quality data, allowing for easy replication and consistency later on. It is a need for the person responsible for the maintaining process that the procedure and frequency of data cleaning be standardized taking into account the most often used data.

Step 6: Every week or month, a specific amount of time must be set aside to closely examine the flaws, approaches that are performing effectively, areas for improvement, and errors and glitches that are occurring.

1.2.4.2.4 Essential Elements for Quality Data

Establishing the standard of information necessitates an examination of its qualities, followed by a ranking of those traits in terms of their value and applicability in the company. The following are the qualities that must be present in high-quality data:

Authenticity: The amount to which the data conforms to stated business restrictions and regulations.

Correctness: The information must be able to depict the real and best values.

Exactness: The amount to which you are familiar with all of the required info.

Data synchronizes inside the same database or across distinct databases and is referred to as data synchronization. Homogeneity refers to how closely the data adheres to the same measurement units.

1.2.4.2.5 Uses of Data Cleaning

Obtaining clean and quality data will almost certainly boost overall productivity and provide high-quality data for rapid and accurate decision-making. Because many sources of data are involved, errors are eliminated to ensure seamless operation. Clients will be happier and more satisfied if there are few to no errors, and employees will be less stressed. It's easier to troubleshoot inaccurate data for future applications when you keep track of problems and have a higher standard of reporting and pinpointing the source of issues.

1.2.4.3 Data Exploratory

Exploratory data analysis (EDA) is used by data scientists to assess, study, and describe data sets' primary attributes, often employing data visualization tools. It makes it easier for data scientists to uncover patterns, test hypotheses, and check assumptions by assisting them in determining how to best manipulate data sources to achieve the answers they require.

EDA is commonly utilized and see what stats can reveal outside of conventional planning or hypothesis testing procedures, as well as to learn more about data set factors and their interactions. It could also help you figure out if the analysis tools you're considering for data analysis are appropriate.

1.2.4.3.1 Requirement of Exploratory Data Analysis

The main purpose of EDA is to help with data analysis before jumping to any conclusions. It can help in the spotting of obvious mistakes, providing a better understanding of data trends, detecting extremes or unforeseen issues, and discovering fascinating correlations between variables.

Data scientists can employ an investigative study to determine whether the results produced are reliable and acceptable for any specific business goals and outcomes. EDA also supports stakeholders by ensuring that the right questions are being asked. It can address questions about standard deviations, predictor variables, and error bars.

1.2.4.4 Data Visualization

The graphical depiction of information and data is known as data visualization. Data visualization tools make it easy to examine and comprehend trends, outliers, and patterns in data by employing visual elements like charts, graphs, and maps. Data visualization tools and technologies are critical in the Big Data environment for analyzing enormous volumes of data and making information choices.

1.2.4.4.1 Essential of Data Visualization

It's difficult to imagine a career that doesn't result in better data understanding. Analyzing data is beneficial to every STEM profession, as well as fields such as administration, business, advertising, olden times, consumer products, skilled trades, education, and sports. Even though we'll always have rhapsodies about data visualization, there are clear practices. Because visualization is so common, it's still one of the most valuable career talents to learn. The more you can graphically communicate your arguments, whether it's in a display or a flash deck, the more effectively you could use that data.

1.2.5 Introduction to Machine Learning

ML is a field that straddles computer science, engineering, and statistics, with applications in a variety of fields. It can be used in a variety of sectors, from politics to geosciences, as you'll discover later. It's a versatile tool that can be used to solve a variety of issues. ML techniques can be used in any industry that requires data interpretation and action. Statistics are used in ML. To the majority of individuals, statistics is a technical topic employed by businesses to exaggerate the value of a product. We have adequate computer resources to model the problem properly. We need statistics to solve these issues. Human motivation, for example, is a challenge that is now impossible to model. Engineering is the application of science to the solution of a problem. We're used to solving deterministic problems in engineering, wherein every answer always solves the problem. If we're writing software to control a vending machine, it needs to work all of the time, regardless of how much money is put in or how many buttons are touched. There are numerous problems for which there is no predictable solution. To put it another way, we don't know enough about the problem or don't have enough computational capacity to simulate it properly. We need statistics to solve these issues. Concept of motivation, for example, is a challenge that is now impossible to model.

1.2.5.1 Necessity of Machine Learning

The progress of ML has been fueled by the virtually infinite amount of data available, affordable data storage, and the development of less expensive and more powerful computing. Many sectors are now building more powerful models that can analyze more and more complicated data while delivering faster and higher accuracy on massive sizes. Organizations can use ML tools to detect profitable possibilities and potential dangers more quickly.

ML's practical applications produce business outcomes that can have a significant impact on a firm's profitability. Innovative approaches in the discipline are continually advancing, allowing for practically unlimited applications of ML. Companies that rely on large amounts of data and require a system to evaluate it quickly and accurately have adopted ML as the most effective technique to create models, strategize, and plan.

1.2.5.2 Applications of Machine Learning

1.2.5.2.1 Machine Learning Usage in Industries

a. **Healthcare:** Portable Iot sensors that track all from heart rates and steps taken to oxygen and serum glucose and even sleep habits have created a large amount of data that allows doctors to examine their patients' health in real time. A new ML algorithm detects malignant tumors on mammograms, identifies skin cancer, and analyzes retinal images to identify diabetic retinopathy, to name a few examples.
b. **Government:** Government officials can utilize data to foresee possible future scenarios and adjust to fast-changing events using ML systems. ML can aid in filtering citizen policy data and building of automatic alerts for the executives about the rules and regulation that they need to remember.
c. **Cybersecurity** and cyber intelligence assist in counter-terrorism activities, the optimization of operational capability, warehouse management, and forecasting, as well as the reduction of failure rates. The healthcare industry has ten more uses for ML, according to a recent article.
d. **Sales and Marketing:** The ML algorithm learns the data patterns from the historical sales data and uses them to predict the future sales and even gives insights to increase the sales.

Many firms have effectively adopted artificial intelligence (AI) and ML to raise and enhance customer satisfaction by over 10%, demonstrating that ML is transforming the marketing field. According to Forbes, "57 percent of enterprise leaders feel that increasing customer experiences and support will be the most critical growth benefit of AI and ML."

E-commerce and social networking sites employ ML to evaluate your purchase and search history and offer recommendations for other goods to buy based on your previous purchases. Most researchers estimate that AI and ML will drive the future of retail as systems improve their ability to capture, analyze, and use data to tailor customers' customer experience and generate personalized, appropriate marketing strategies.

Shipping: Profitability in this industry depends on efficiency and precision, as well as the capacity to forecast and manage possible difficulties. Data analysis and modeling capabilities of ML are ideal for businesses in the delivery, public transportation, and freight transportation industries. ML is a vital component of supply chain management since it employs algorithms to uncover elements that favorably and negatively affect the success of a supply chain. During shipping, ML enables scheduling algorithms to optimize provider selection, rating, routing, and quality control operations, saving money and increasing efficiency. Because of its ability to evaluate thousands of data points at once and apply algorithms faster than a human, ML can address issues which are yet to be defined.

1.2.5.2 Relationship between Machine Learning and Data Analysis

The opportunity to comprehend and identify prospects and customers rises in tandem with the growth of consumer data.

- Businesses must define their strategy strategically to capitalize on this data. After all, simply having the information isn't enough to:
  - interpret and comprehend the story being told,
  - assess whether information would be most useful to which client, and
  - inspire in employees a culture of data discovery, especially if acting on hunches is a habit.

  In this way, analytics software that naturally fosters data-driven decision-making gives you a leg up on the competition.

The introduction of AI analytics has altered the conversation's foundation. Analytics tools are not only enablers of data gathering but also able to complete the real labor that was once unique to people, thanks to AI's automation and augmentation capabilities.

Business leaders recognize the importance of data customized for each function, as well as the role analytics tools play in the entire workforce of using that data.

1.2.5.3 Necessity of Probability for Machine Learning

It's possible to infer that probability is required to effectively complete an ML predictive modeling project. The technique of developing prediction models from contradictory data is known as ML. Uncertainty typically works with incomplete or inaccurate facts.

Though inconsistency is critical to ML, it is now one of the aspects that beginners, notably those with a programming experience, find the most challenging.

There are three main sources of ambiguity in ML: noisy data, poor penetration of the issue region, and inappropriate algorithms.

Nonetheless, we may use probability methods to handle ambiguity. To control the ambiguity we see in each project as ML practitioners, we need to know probability.

Machine learning is built on the foundation of probability.

- Model-based approaches must forecast the likelihood of belonging to a given class.
- Probability is used to build algorithms (e.g. Naive Bayes).
- Probability will be used by learning algorithms to make decisions (e.g. information gain).
- Probability is the foundation of subfields of study (e.g. Bayesian networks).
- Probability frameworks are used to train algorithms (e.g. maximum likelihood).
- Probabilistic loss functions are used to fit models (e.g. log loss and cross entropy).
- Probability is used to configure model hyperparameters (e.g. Bayesian optimization).
- Design skill is assessed using probabilistic methods.

1.2.5.4 Three Types of Machine Learning Algorithm

Commonly used ML algorithms are given specific attention. The algorithms studied include linear regression, logistic regression, Naive Bayes, kNN, random forest, and others. The basic three types of ML algorithms are as follows.

1.2.5.4.1 Supervised Learning

This algorithm is made up of a target/result variable (or dependent variable) that must be estimated from a set of predictor variables (independent variables). We create a function that maps inputs to desired outputs using this set of variables. The model is trained until it accomplishes the appropriate level of accuracy on the data set. Regression, decision tree, random forest, KNN, logistic regression, and others are examples of supervised learning.

1.2.5.4.2 Unsupervised Learning

This algorithm is made up of a target/result variable (or dependent variable) that must be estimated from a set of predictor variables (independent variables). We create a function that maps inputs to desired outputs using this set of variables. The model is trained until it accomplishes the appropriate level of accuracy on the data set. Regression, decision tree, random forest, KNN, logistic regression, and others are examples of supervised learning.

1.2.5.4.3 Reinforcement Learning

The machine is taught to make certain decisions using this algorithm. It works like this: the machine is deemed to occur where it must constantly train itself through trial and error. This computer adapts from its previous experiences and attempts to gather the most relevant information in order to make appropriate business decisions.

1.3 Conclusion

ML is a type of data analysis that facilitates the creation of analytical models. It helps computers to discover underlying truths without becoming specifically instructed wherever to seek through using algorithms that continuously help in decision-making.

ML now is not the same as ML in the past, thanks to advances in computing technology. Whereas many ML techniques have indeed been known for a while, the capacity to apply difficult mathematical computations to large amounts of data automatically—again and over, quicker and faster—is a relatively new phenomenon.

Chapter 2

Role of Machine Learning in Promoting Sustainability

Muneza Kagzi

Contents

DOI: 10.4324/9781003206316-2

2.1 Introduction

There is a universal agreement that proactive measures are needed to save the planet (Pérez-Ortiz, de La Paz-Marín, Gutiérrez, & Hervás-Martínez, 2014). The importance of sustainability has increased tremendously on the back of the rising number of natural calamities inflicting the planet. The central idea of sustainability is based on the fact that human consumption is far exceeding what the planet is able to reproduce. This rising difference between reproduction and consumption is a cause of worry, shifting focus to sustainability as a means to bridge the gap. Sustainability has attracted much attention from researchers, policy-makers, consumers, and investors across the globe (Zhao, Liu, Zhang, & Fu, 2020). Elkington (1998) conceptualized an approach to achieving sustainability by following the triple bottom lines that represented the economic, environmental, and social criteria of corporate performances. To implement sustainable development, researchers have recommended the implementation of various machine learning (ML) techniques.

The wider application of ML has gained momentum in recent times. ML seeks to imitate human learning by using input data, algorithms, and statistical methods to make predictions or decisions (Fahdi, Elkhechafi, & Hachimi, 2021). ML techniques identify the pattern in data and produce an estimated outcome. These techniques help to identify the problem of resource consumption in the domain of sustainability and subsequently use data and insights to design optimized resource consumption solutions. This conceptual chapter conducts an in-depth analysis of existing research on ML and how the insights can be leveraged to further the concept of sustainability with respect to the three P's—namely, people, profit, and planet.

2.2 Machine Learning for Planet

Environmental sustainability is very crucial for the survival of the planet (Elkington, 1998). Enterprises need to reduce their carbon footprint in their business activities and make them more sustainable. An extensive review of the literature throws up some interesting insights into how enterprises can reduce their carbon footprint and follow sustainable development practices in the interest of the planet.

2.2.1 Machine Learning for Planning and Promoting Sustainable Cities

Urbanization is one of the driving forces of economic growth. However, this continuous process of urbanization has been at the cost of the environment, and today, there is more need than ever to follow sustainable development practices. Li et al. (2019) proposed the use of an ML application, the agent-based modeling with embedding for geo-simulation, to create sustainable cities and communities. This methodology uses past location data to make new location decisions and shape sustainable urban development.

Lan, Zheng, and Li (2021) combined neural networks and deep learning and in urban planning with a view to incorporate principles of sustainability.

2.2.2 Machine Learning for Waste Management

The management of municipal solid waste (MSW) is a crucial aspect of sustainable development of the urban space. However, MSW sometimes does not get the attention it deserves mainly due to the lack of adequate waste management infrastructure. To manage MSW effectively, it is important to accurately forecast the MSW composition and generation rate. However, forecasting MSW can be challenging in the context of low-income countries, where data is either unavailable or unreliable. Research (Ayeleru, Fajimi, Oboirien, & Olubambi, 2021) recommended two ML techniques—artificial neural network (ANN) and supported vector machine (SVM) for resolving the issue in the context of the City of Johannesburg, South Africa, to forecast the quantity of MSW generation. Accurate MSW forecasting, therefore, can go a long way to forecast MSW, thus enabling authorities to plan for adequate waste management and infrastructure.

2.2.3 Machine Learning for Clean Energy

ML can be applied to promote clean energy generation for the benefit of the planet. Geothermal power plants are a means to generate clean energy. Geothermal energy refers to the energy that is generated from the rocks and fluids found beneath the earth's crust, as far down to the hot molten rock, magma (Coro & Trumpy, 2020). However, the effectiveness of the geothermal power plants is dependent on the area where it is located and also on several environmental factors. ML can be used for renewable energy predictions. The data-driven ML models can be used to create suitability maps of geothermal sites. Thus, ML can help to identify special geographies for geothermal power plant installations. The assessment of suitable geographies for geothermal suitability involves high costs, invasive inspections, and legal permissions. Research proposes the Maximum Entropy (MaxEnt), an ecological modeling ML model, to create suitability maps for the setting up of geothermal power plants (Coro & Trumpy, 2020). The MaxEnt algorithm model can be trained with environmental data and be potentially correlated with geothermal site suitability and geothermal plant operations. Such data-driven deep learning models can help to make renewable-energy predictions, thus saving time, money, effort, and other valuable resources. ML, in the context of clean energy, can, therefore, equip policy-makers, geothermal energy industry operators, geologists, and territory citizens with the relevant data to assess suitable sites and plan for future geothermal power plants.

2.3 Machine Learning for People

Sustainable development is crucial for the health and well-being of all people associated with an enterprise, be it the employees or the external communities and stakeholders, such as suppliers, customers, and shareholders. Existing research testifies to the following ways in which ML helps to improve well-being of the people.

2.3.1 Machine Learning for Assisting Green Customers

ML promotes responsible production and consumption. Consumers today are more environmentally aware and prone to purchase products or services that have low environmental impact. More and more brands are now showcasing their green credentials to build credibility with their customers. As a case point, the hospitality and tourism sector is increasingly using ML for sustainable business practices (Khanra, Dhir, Kaur, & Mäntymäki, 2021). For instance, the popular online travel company with user-generated content feeds consumer feedback data in its ML program to create a framework for recommending green hotels. The ML methods for such predictions include Self-Organizing Map (SOM), Adaptive Neuro-Fuzzy Inference System (ANFIS), Higher-Order SVD (HOSVD) technique, and Classification And Regression Tree (CART) (Nilashi et al., 2019).

2.3.2 Machine Learning to Safeguard the Interest of Shareholders

ML can promote shareholders' interest by providing information on whether or not the corporates are following ethical practices. Researchers have suggested the application of random forest (RF) model to fulfill the variable importance ranking and corporate misconduct prediction (Wang, Asghari, Hsu, Lee, & Chen, 2020). This model was used in the context of China. It took into consideration observations from construction companies in 2000–2018. The RF model identified the key variables of corporate governance. These variables could be used to forecast corporate misconduct. Such findings can help policy-makers to improve decision-making and better regulate malpractices that adversely impact the environment. Further, they can help regulators in timely identifying violating companies and implement proactive interventions.

2.3.3 Machine Learning for e-Government Services for Citizens

Prior studies have highlighted that effective use of technology may help government authorities to better serve their citizens (Khanra, Joseph, & Ruparel, 2019). For instance, Khanra and Joseph (2020a) developed a mathematical model for a smooth delivery of e-Government services. Applications of ML can help to automatically classify civic queries and ensure their faster disposal (Kim, Yoo, Park, Lee, & Kim, 2021). Kim et al. (2021) developed a dynamic topic model to identify the associations among the keywords in civic complaints. The model processes words in text documents and classifies queries following the Word2vec approach to provide data-driven recommendations (Kim et al., 2021). However, demographic characteristics and service quality can impact adoption and dissemination of e-government services (Khanra & Joseph, 2017), as also English proficiency and digital divide (Khanra & Joseph, 2019). Therefore, Kim et al. (2021) tested their ML model with a large data pool of 1,60,316 civic complaints raised between 2006 and 2017 in Seoul, South Korea. Furthermore, Khanra and Joseph (2020b) developed an assessment framework for dissemination of effective e-Government services.

2.4 Machine Learning for Profit

Sustainable development refers to the process of generating economic profit while balancing the environmental and social aspects. To this extent, prior studies recommend the following solutions.

2.4.1 Machine Learning at the Company Level

To generate profit, companies can redesign their existing processes to make them more cost-effective—as a case point, redesign existing concrete machines used in the cement industry. Concrete is largely used in the construction industry creating environmental challenges such as energy consumption, depletion of natural resources, and greenhouse gas emission (Naseri, Jahanbakhsh, Hosseini, & Nejad, 2020). Therefore, there is a need to redesign eco-friendly versions of these concrete machines that are cost-effective, have compressive strength, minimize energy and resource consumption, and reduce greenhouse gas emissions (Naseri et al., 2020). To design the green concrete machine, the most efficient objective function is used specifying various sustainability parameters (Naseri et al., 2020).

Researchers have proposed ML techniques that can predict important mechanical characteristics of concrete, for instance, its compressive strength to determine its quality (Naseri et al., 2020). The higher the compressive strength of the concrete machine, less the amount of energy consumption and material used, less the amount of greenhouse gas (embodied CO2) emitted, and ultimately less the budget required for manufacturing. To forecast compressive strength of eco-friendly concrete, six ML algorithms were proposed (Naseri et al., 2020). These include the soccer league competition algorithm, water cycle algorithm (WCA), ANN, genetic algorithm (GA), regression, and support vector machine (SWM) (Naseri et al., 2020). Of these, results show that WCA is the most efficient model for predicting an eco-friendly concrete machine given that it considers mean absolute error and coefficient of determination. WCA helps to optimize the use of resources in designing an eco-friendly concrete machine.

Accordingly, the cost, compressive strength, and environmental impacts, including embodied carbon dioxide emission, and energy and material requirements, are considered as sustainability attributes (Naseri et al., 2020). To integrate their influence on mixture designing, six objective functions are applied. These include the fractional-based sustainable objective function, linear form sustainable objective functions, and distances of sustainability parameters among others. As per the name of water cycle algorithm, it is based on the idea of the process of river flowing into sea from various places. In the context of designing eco-friendly concrete machines, optimization plays an important role in utilizing multiple resources and objective functions.

2.4.2 Machine Learning at the Industry Level

Mele and Magazzino (2020) have recommended applications of ML to promote sustainability in the Chinese steel industry. An ML model named long short-term memory may be used to balance economic growth and environmental impact to sustain the steel industry in the long run (Mele & Magazzino, 2020). The concept of memory featuring the data may be utilized for prediction using gates (Mele & Magazzino, 2020). This model may examine the relationships among the three important variables, namely production in the iron and steel industries, air pollution, and economic development (Mele & Magazzino, 2020). The network of long short-term memory is based on the recursive neural network (Mele & Magazzino, 2020). Therefore, application of ML may better predict the outcome of steel production on the economy and environment, helping to balance production and its environment impact (Mele & Magazzino, 2020). Furthermore, as more and more manufacturing firms strive to servitize, manufacturing sectors may embrace greater use of ML to reduce resource consumption (Khanra, Dhir, Parida, & Kohtamäki, 2021).

2.4.3 Machine Learning at the Country Level

The sugarcane industry's expansion in Brazil led to deforestation in the Amazon rainforest, leading to climate change and water and energy insecurities. This built the case for the industry to adopt sustainable practices. In this context, the researchers have used ML techniques to identify the commitment of an enterprise to sustainability. The Latent Dirichlet Allocation (LDA) algorithm can be used to extract topics/themes from a large number of documents, such as websites, sustainability reports, and flyers (Benites-Lazaro, Giatti, & Giarolla, 2018). LDA can analyze the data and predict if the company follows sustainable practices. An LDA analysis in the context of the Brazil sugar industry revealed 36 key themes that demonstrated the initiatives it has taken to promote sustainability. These include formulating standards and codes of conduct and implementing various CSR programs. Few of the themes included bioelectricity, agro-environmental, consumption, biofuels, emission, certification, UNICA, standard, code, climate, labor, committee, report, education, financial, flex-fuel, position, sustainable, health, trade, industry, job, land, voluntary, local, life, logistic, technology, model, preservation, production, program, riparian, supplier, tariff, and water.

2.5 Discussion and Conclusion

This conceptual chapter examines the role of ML in achieving sustainability. It aptly demonstrates with examples how ML can contribute to sustainable development with impacts at three levels, that is, people, profit, and planet. The study also develops a framework, as presented in Figure 2.1, to posit that the application of ML tools and techniques will benefit society, the environment, and the industries. This framework also outlines the specific ML techniques and the context of their application.

For all data-driven predictions related to a sustainable planet, research recommends the use of agent-based model (Li et al., 2019), deep neural network, and the random forest model (Lan et al., 2021). Urbanization is witnessing increased growth on the back of economic growth. Location planning using ML will enable authorities to develop sustainable urban solutions. Artificial neural network and SVM methodology (Ayeleru et al., 2021) have found use in designing effective waste management solutions. Maximum entropy (Coro & Trumpy, 2020) has found use in mapping locations for geo-thermal plants, which can serve as sources for clean energy. ML methods such as SOM, ANFIS, HOSVD technique, and CART (Nilashi et al., 2019) are being extensively used in the tourism and hospitality sectors,

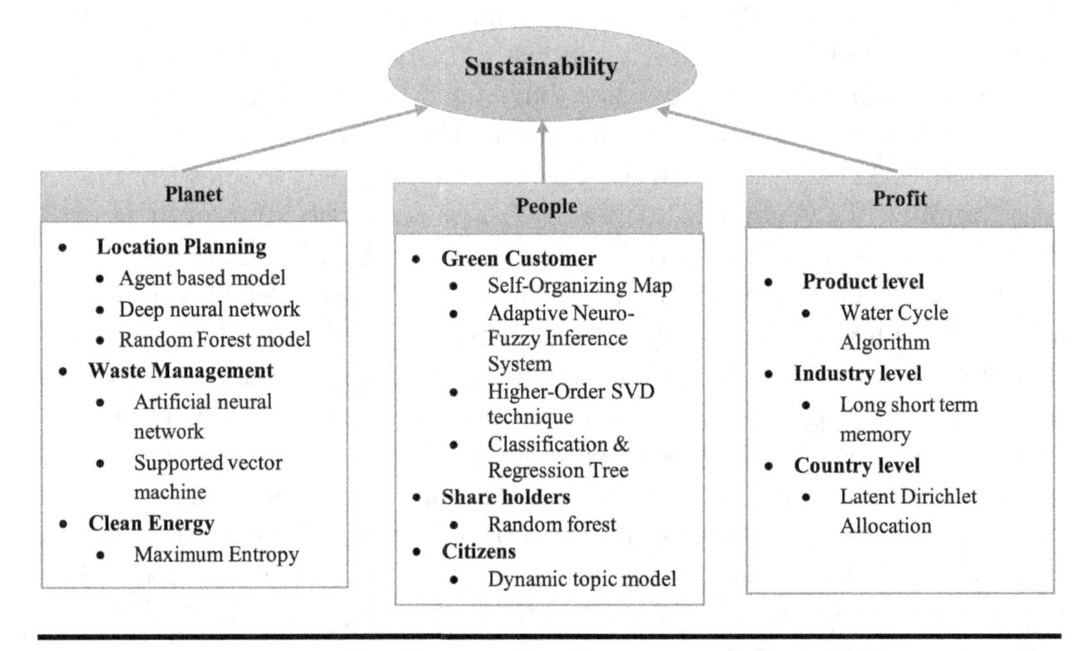

Figure 2.1 Application of machine learning for promoting sustainability.

for instance, to recommend green hotel accommodations using user-generated content as data, as in the case of the e-platform, TripAdvisor.

ML can also promote the well-being of people within and outside an organization. Fuzzy set theory and ensemble learning technology can help to manage green suppliers (Wu, Lin, Barnes, & Zhang, 2020). On the other hand, SOM, ANFIS, HOSVD, and CART have found use in promoting green customers (Nilashi et al., 2019). Furthermore, random forest methodology can help to predict unethical practices in a corporate setting, thus helping to protect shareholders' rights.

To promote profitability, ML can be implemented at three levels, namely, product, industry, and country. As a case point, WCA can be used to design a sustainable cement concrete machine (Naseri et al., 2020). Similarly, LDA can be implemented at the country level to promote sustainability in the sugarcane industry (Benites-Lazaro et al., 2018). Long short-term memory (Mele & Magazzino, 2020) can be utilized to sustain the steel industry in the long run. The concept of memory featuring data can be used to forecast the output of the steel industry and measure the impact it has on the environment and economy. From the previous discussion, it becomes clear that various ML techniques can be used to combat climate change and promote green practices in all aspects of our activities. While ML is still at the nascent stage, given the success it has met with, its use will grow all the more as it evolves.

References

Ayeleru, O. O., Fajimi, L. I., Oboirien, B. O., & Olubambi, P. A. (2021). Forecasting municipal solid waste quantity using artificial neural network and supported vector machine techniques: A case study of Johannesburg, South Africa. *Journal of Cleaner Production*, 289, 125671.

Benites-Lazaro, L. L., Giatti, L., & Giarolla, A. (2018). Sustainability and governance of sugarcane ethanol companies in Brazil: Topic modeling analysis of CSR reporting. *Journal of Cleaner Production*, 197, 583–591.

Coro, G., & Trumpy, E. (2020). Predicting geographical suitability of geothermal power plants. *Journal of Cleaner Production*, 267, 121874.

Elkington, J. (1998). Partnerships from cannibals with forks: The triple bottom line of 21st-century business. *Environmental Quality Management*, 8(1), 37–51.

Fahdi, S., Elkhechafi, M., & Hachimi, H. (2021). Machine learning for cleaner production in port of Casablanca. *Journal of Cleaner Production*, 294, 126269.

Khanra, S., Dhir, A., Kaur, P., & Mäntymäki, M. (2021). Bibliometric analysis and literature review of ecotourism: Toward sustainable development. *Tourism Management Perspectives*, 37, 100777.

Khanra, S., Dhir, A., Parida, V., & Kohtamäki, M. (2021). Servitization research: A review and bibliometric analysis of past achievements and future promises. *Journal of Business Research*, 131, 151–166.

Khanra, S., & Joseph, R. P. (2017). Adoption and diffusion of e-government services: The impact of demography and service quality. In *Proceedings of the 10th International Conference on Theory and Practice of Electronic Governance* (pp. 602–605). ACM Digital Library, New York, NY, USA.

Khanra, S., & Joseph, R. P. (2019). Adoption of e-governance: The mediating role of language proficiency and digital divide in an emerging market context. *Transforming Government: People, Process and Policy*, 13(2), 122–142.

Khanra, S., & Joseph, R. P. (2020a). Orchestration of an e-government network: Capturing the dynamics of e-government service delivery through theoretical analysis and mathematical forecasting. In *International Working Conference on Transfer and Diffusion of IT* (pp. 219–229). Springer, Cham.

Khanra, S., & Joseph, R. P. (2020b). Development and validation of an assessment framework for e-government services. In *International Conference on Electronic Governance and Open Society: Challenges in Eurasia* (pp. 27–41). Springer, Cham.

Khanra, S., Joseph, R. P., & Ruparel, N. (2019). Dynamism of an e-government network in delivering public services. In *Academy of Management Global Proceedings* (p. 376). Academy of Management,USA.

Kim, B., Yoo, M., Park, K. C., Lee, K. R., & Kim, J. H. (2021). A value of civic voices for smart city: A big data analysis of civic queries posed by Seoul citizens. *Cities*, 108, 102941.

Lan, H., Zheng, P., & Li, Z. (2021). Constructing urban sprawl measurement system of the Yangtze River economic belt zone for healthier lives and social changes in sustainable cities. *Technological Forecasting and Social Change*, 165, 120569.

Li, F., Xie, Z., Clarke, K. C., Li, M., Chen, H., Liang, J., & Chen, Z. (2019). An agent-based procedure with an embedded agent learning model for residential land growth simulation: The case study of Nanjing, China. *Cities*, 88, 155–165.

Mele, M., & Magazzino, C. (2020). A machine learning analysis of the relationship among iron and steel industries, air pollution, and economic growth in China. *Journal of Cleaner Production*, 277, 123293.

Naseri, H., Jahanbakhsh, H., Hosseini, P., & Nejad, F. M. (2020). Designing sustainable concrete mixture by developing a new machine learning technique. *Journal of Cleaner Production*, 258, 120578.

Nilashi, M., Ahani, A., Esfahani, M. D., Yadegaridehkordi, E., Samad, S., Ibrahim, O., Sharef, N. M., & Akbari, E. (2019). Preference learning for eco-friendly hotels recommendation: A multi-criteria collaborative filtering approach. *Journal of Cleaner Production*, 215, 767–783.

Pérez-Ortiz, M., de La Paz-Marín, M., Gutiérrez, P. A., & Hervás-Martínez, C. (2014). Classification of EU countries' progress towards sustainable development based on ordinal regression techniques. *Knowledge-Based Systems*, 66, 178–189.

Wang, R., Asghari, V., Hsu, S. C., Lee, C. J., & Chen, J. H. (2020). Detecting corporate misconduct through random forest in China's construction industry. *Journal of Cleaner Production*, 268, 122266.

Wu, C., Lin, C., Barnes, D., & Zhang, Y. (2020). Partner selection in sustainable supply chains: A fuzzy ensemble learning model. *Journal of Cleaner Production*, 275, 123165.

Zhao, S., Liu, Y., Zhang, R., & Fu, B. (2020). China's population spatialization based on three machine learning models. *Journal of Cleaner Production*, 256, 120644.

Chapter 3

Addressing the Utilization of Popular Regression Models in Business Applications

Meganathan Kumar Satheesh and Korupalli V. Rajesh Kumar

Contents

DOI: 10.4324/9781003206316-3

3.1 Introduction

Globalization has made companies change the strategy, both internal and external, for maintaining a good relationship with business partners and customers (Bergeron, Raymond, & Rivard, 2004). This makes the organization do a lot of prediction and forecasting to achieve the goal. However, a bad prediction of the business failure will affect the success of the business due to inappropriate decisions and will reduce profit for the stakeholders due to a fall in the market value of the firm (Sofie & Hubert, 2004). Prediction models are also used to understand the connections between the buyers and sellers, paving the way for comparing the various predictive models to get higher accuracy (Zuo, Kajikawa, & Mori, 2016).

Statistical models can be effective if their accuracy is higher and the model outcome is closer to reality (Palasca, 2012). Regression is one of the techniques used to determine the relationship between independent and dependent variables (Hopkins & Ferguson, 2014; Hosmer Jr, Lemeshow, & Sturdivant, 2013), which may be either linear or non-linear used for forecasting (Bansal, Kauffman, & Weitz, 1993). A regression model is performed to select the best independent variables along with their correlation and magnitude that have a direct effect on the dependent variable (Larasati, DeYong, & Slevitch, 2011). There is a difference between correlation that may be either positive or negative, but it does not indicate one is the cause for the other and causality that is a confirmation of causal relationship based on the causal theory (Rubinfeld, 2011). A hypothesis test can be performed to substantiate the proposed theory using cross-sectional analysis when the data is a sample from the population at a specific point of time and time series analysis when the data is collected from different periods (Rubinfeld, 2011). The discussion of different regression models is in Section 3.2 and the business applications of the regression model will be discussed in Section 3.3. The conclusion will be discussed in Section 3.4.

3.2 Popular Regression Models

3.2.1 Simple Linear Regression

Simple linear regression is the technique used to find the prediction and correlation between an independent variable and a dependent variable (Bayaga, 2010; Bolton, 2009; Omay, 2010), which is also used in machine learning (ML) to get the output from the linear form of connections using input variables (Buteneers, Caluwaerts, Dambre, Verstraeten, & Schrauwen, 2013). The assumptions of linear regression are as follows: the independent and dependent variables should have a linear relationship; a variance of residuals should be constant; values (residual) should be independent; and the distribution should follow the normal distribution (Bolton, 2009).

3.2.2 Multiple Linear Regression

Assumptions of multiple linear regression will be the same as simple linear regression (Omay, 2010) along with avoidance of multi-collinearity that will increase the chance of type II error (Qasim, Månsson, & Kibria, 2021), among the independent variables (Detienne, Detienne, & Joshi, 2003). Incorporating all the dependent variables will give the model strength to predict the result precisely; failing to do so will cause bias in the model affecting the statistical result (Rubinfeld, 2011).

3.2.3 Logistic Regression

Logistic regression, which is an extension of logit transformation (Omay, 2010), is different from linear regression in terms of dependent variable outcome (Hosmer Jr et al., 2013; Omay, 2010) that is either in binary form or dichotomous form(Bolton, 2009). If the regression is linear regression, the dependent variable outcome will be in continuous form. On the other hand, logistic regression will have a discrete or categorical outcome for the dependent variable (Hosmer Jr et al., 2013; Omay, 2010). Binomial logistic regression is used in the case of two categorical outcomes for a dependent variable, whereas more than two categorical outcomes for a dependent variable can be handled by multi-nomial logistic regression (Bayaga, 2010). Here, there is no need for the independent variables to be in the normal distribution and for both variables to be in a linear relationship, paving the way for more flexibility when compared to multiple linear regression (Bayaga, 2010; Omay, 2010). An ordinal logistic regression model is preferred when the dependent variable is in the ordinal form (rank order), which does not help deal with multi-collinearity between independent variables (Larasati et al., 2011).

3.2.4 Ridge Regression

Ridge regression, also known as a curative model of linear regression (Kibria, Månsson, & Shukur, 2011), is useful in handling multi-collinearity between the independent variables with the help of ridge parameter and identity matrix and is given a higher performance in results than ordinary least-squares method particularly in dealing with high correlation between independent variables (Bager, Roman, Algelidh, & Mohammed, 2017). It incorporates an extra element to the loss function, which is useful in tuning the regularization parameter of ridge regression for eliminating the overfitting problem (Buteneers et al., 2013). Besides, a likelihood function is improved by penalizing only the coefficients of the independent variables but not the intercepts (Pereira, Basto, & Silva, 2015). However, the ridge regression with this penalty will make the coefficients nearly close to zero but not exactly zero creating a hindrance in selecting the predictors in the model (Pereira et al., 2015).

3.2.5 Lasso Regression

The problem in decreasing the independent variables by the ridge regression will be solved by the lasso regression as the introduced penalty in the lasso regression acts not only to direct the coefficients to zero but also to select the correct independent variables in the model (Pereira et al., 2015).

3.2.6 Random Forest Regression

Random forest is an amalgamation of various trees in which each tree subset of the observation is chosen randomly (Grömping, 2009). Random forest regression is used for Boston housing price prediction (Liaw & Wiener, 2002). Because the overall prediction is based on the prediction of an individual tree that delivers a multidimensional step function, the outcome of the overall prediction will be smooth (Grömping, 2009).

3.2.7 Support Vector Regression

Support vector regression is based on learning theory (statistics), which is used for the classification of objects as well as optimal characters (Smola & Scholkopf, 2004). This gives the best result when compared to another model like feature space representation on Boston house price estimation (Drucker, Burges, Kaufman, Smola, & Vapnik, 1996).

3.2.8 Quantile Regression

Quantile regression, which is an extension of quantile function (conditional), is used for the estimation of a conditional-based model (Koenker & Hallock, 2001). The word quantile means a sample is divided into equal sized and sub groups, which means 25%, 50%, and 75% are the sample quantiles (Yu, Lu, & Stander, 2003) that can be used to observe extreme values of the samples (Jareño, Ferrer, & Miroslavova, 2016). Linear regression will use only the average relationship between dependent and independent variables, whereas the quantile regression will give a clear picture of the relationship between both variables by plotting quantile regression curves (Yu et al., 2003). The coefficients obtained by quantile regression will be distributed by the outliers (Jareño et al., 2016) due to the usage of a weighted sum of absolute deviation, and the estimators of quantile regression will deliver an effective performance than ordinary least square due to error terms that are not in the normal distribution (Hung, Shang, & Wang, 2010).

3.2.9 Limitations of Regressions

Multiple linear regression has a severe drawback in handling the nonlinearity problem (Detienne et al., 2003). The sophistication of the regression model will be increased if the correlation is high between independent variables, leading to the elimination of some important independent variables due to multi-collinearity and the addition of some other unwanted independent variables (Pereira et al., 2015). The assumption made in modeling the regression is based on the researcher's choice that will lead to affect the decision made from the regression results (Detienne et al., 2003). Also, adjusting the regression line from the least square value will affect the assumption from which the regression model is built (Bolton, 2009).

Besides, the elimination of coefficients that is nearly equal to zero will not give any contribution to the regression model, so the improvisation of data is not there (Detienne et al., 2003). Overfitting will be a problem when the number of samples (rows) is lower than the total number of independent variables (Pereira et al., 2015). If the variables are correlated more strongly, even more data will not be helpful (Detienne et al., 2003). The ridge regression has a disadvantage that standard errors lack, which makes the predicted coefficients cannot be tested for their significance (Annaert, Claes, Ceuster, & Zhang, 2013). When it comes to outliers in the data, linear regression is not effective in dealing with those outliers (Detienne et al., 2003).

3.3 Applications of Regression Models in Four Domains of the Business

3.3.1 Finance

3.3.1.1 Credit Score

Simple linear regression is used to find the credit score (Bolton, 2009) that has been evaluated for not only individuals (customers) but also businesses (Bensic, Sarlija, & Zekic-Susac, 2006). This credit score is effectively useful to understand the potential of the borrower in terms of repayment. Even if the credit score evaluation is for business, the owner of the business will be evaluated for the repayment of borrowed loan amount by measuring his/her credit score (Bensic et al., 2006) where the logistic regression is used for deciding whether the credit risk is positive or negative (Satheesh & Nagaraj, 2021). Bolton (2009) has collected data from the bank that was used to form various clustering after eliminating the non-significant variables. Then, stepwise logistic regression was used to increase the potential of the model in predicting the credit score (Bolton, 2009).

3.3.1.2 Risk Analysis

Awareness, identification, tracking, and palliation of risk are required to develop and analyze the risk analysis method. A risk analysis model can be built from the relationship between the independent and the dependent variables. Multiple linear regression has given a better accuracy rate of 60.5% in determining the likelihood of the risk with the help of significant predicting variables (Bayaga, 2010).

3.3.1.3 Business Performance

There are two elements that aid in the evaluation of business performance: internal behavior and external environment along with various organizational performance determinants like return on investment, profits earned, size of the company, number of years since establishment, and its position in the association (Wood, 2006). An ordinal logistic regression and a binary logistic regression were used in the United Kingdom small firm data to find the best model that would provide the major factors affecting the firm's performance. And the results showed that the number of active years, promotion procedures, family funds, and struggle in hiring activities were the major determinants that appeared to affect business performance. However, the ordinal logistic regression was delivered a better performance in identifying the significant variables than binary logistic regression (Wood, 2006).

3.3.1.4 Revenue Model

Additional services along with the core service will give more revenue to the organization. For example, motivating passengers to engage in shopping during the waiting time at the airport will increase the revenue for the organization and, at the same time, will improve the traveling experience for the customers (Fasone, Kofler, & Scuderi, 2016). A ridge regression model and a partial least-squares regression model are used to identify the factors in non-aviation services urging to shopping, due to keeping all the independent variables that have a very high correlation with each other (Fasone et al., 2016).

3.3.1.5 Value at Risk

Value at risk model that uses quantile regression approach is an indicator of market risk, which requires banks to inform on a daily basis to develop financial regulations (Yu et al., 2003). Value at risk is based on the financial return that is demonstrated by several quantiles so the quantile regression is the best approach to evaluate it (Yu et al., 2003).

3.3.1.6 Business Failure

Business failure may happen due to various reasons, namely, insolvency, legal problems, or consecutive losses (Pereira et al., 2015). Thus, it is essential to predict corporate failure in advance to make the right decision for preventive measures. Pereira et al. (2015) were used ridge logistic and lasso logistic regression to predict the business failure with another method like the stepwise method, results of the ridge and lasso logistic regression showed lower type I error than the stepwise method.

3.3.2 Marketing

3.3.2.1 Direct Marketing Score Model

Direct marketing is the way of communicating to the potential customers with specified offers by the marketing team to urge them to involve purchase. However, it is not an easy task to do because each communication will cost the company and, at the same time, not every customer is ready to purchase immediately even if the offer is so attractive (Malthouse, 1999). Market segmentation is one of the basic marketing practices to break down the entire market customer into small market customer groups that will be helpful to apply right strategies for enhancing the relationship between the organization and the customers (Chattopadhyay, Dan, Majumdar, & Chakraborty, 2012). Even after segmenting, the targeting should be selected based on the customer's willingness to make a purchase, which can be done using directing marketing score model that has so many explanatory variables (Malthouse, 1999). The ridge regression is used to analyze the performance of direct marketing score because many variables can be included in the model without worry about the multi-collinearity and overfitting assisting marketers to precisely understand the impact of predictive variables on direct marketing (Malthouse, 1999).

3.3.2.2 Family Business Research

The performance of the family business and family ownership handover and the relationship between the involvement of family members in the board and global sales were evaluated with regression analysis (Hopkins & Ferguson,

2014). Moreover, the relationship between ownership of founding family and business performance was also evaluated by multiple linear regression analysis due to the involvement of many independent variables, namely the number of family members involved, incentives of family and outside members, the number of years served in top management positions, and the number of outside members in the board (Hopkins & Ferguson, 2014).

3.3.2.3 Market Consumption and Price Determinants

Accessing individuals to do the study of the market will be difficult, which leads to group formation. Those groups can be divided into high, medium, and low groups and can be studied with the help of quantile regression (Yu et al., 2003). Pricing of a hotel room is influenced by so many factors like marketing tools involved and its strategies, room rates, demand in the market, consumer perception, customer satisfaction, which can be solved and analyzed by quantile regression to get the significant pricing determinants (Hung et al., 2010). Besides, multiple linear regression is used to predict the price of the apartments that have many explanatory variables (Ceh, Kilibarda, Lisec, & Bajat, 2018).

3.3.2.4 Business Growth

Although several incentives are offered by the US government to the US farmers, business strategies—namely, cost reduction, revenue improvement, asset, and financial management—are deployed for maintaining a stable income and equity growth. Those business strategies that are performed in different quantiles can be examined using quantile regression analysis (Hennings & Katchova, 2005). Also, the influence of acquisition performance on the level of company distress and the relationship between monitoring rating and performance and quality along with the duration of work are performed with the help of regression analysis (Aguinis, 1995).

3.3.2.5 Export Market

A company involved in exporting to foreign countries will not have the same resources used by the other export company due to the difference between companies with respect to firm size, availability of skilled labor, research and development facility, etc. (Wagner, 2004). Several organizational hurdles prevent successful exports—namely, poor knowledge of international markets, lack of foreign cultural exposure, and inadequate foreign networks (Bennett, 1997). To find out the relationship between export sales ratio and plant attributes, Wagner (2004) has used the quantile regression.

3.3.3 Human Resource Management

3.3.3.1 Organizational Justice and Employee Commitment

Organizational justice is referred to fairness in decision-making and its perception in organizations, which affects the employee commitment that has a direct connection to the performance of the employees (Jamaludin, 2009). It is classified into three justice: the first one is distributive justice that is perceived understanding of resource distribution, the second one is procedural justice that is related to the compensation assigned to the workload, and the third one is interactional justice that is the interaction between the employees and the organization. There are three types of commitments: an affective commitment that is the emotional and psychological attachment of the employee with the company, normative commitment that is a feeling of responsibility to the organization, and continuance commitment that is a cost-associated consideration of leaving the organization combined with the opportunities outside the organization (Jamaludin, 2009). Jamaludin (2009) has conducted a linear regression model to understand the influence of organizational justice on employee commitment and concluded that interactional justice has a strong impact on affective commitment and distributive justice has no impact on any of the above-mentioned commitments.

3.3.3.2 Employee Selection

Hiring a candidate is very critical in the services industry, where the performance of the employee plays a major role in business success. Many variables are available in the database to be considered to select the right candidate, which has to cluster into four groups: personal variables that include personal details, test variables that include scores of exams and interview, occupational variables that include employment history, and performance variables that include promotion and career score (Sebt & Yousefi, 2015). Sebt and Yousefi (2015) have used stepwise regression analysis to obtain the best predictive variables for the employee selection after performing a decision tree in the data mining process and have concluded that the five best variables are identified from the total of 26 variables.

3.3.3.3 Human Capital Development

To survive in the competitive business world, the organization should have the competitive advantage that can be formed by enhancing the skills of employees by providing proper training and opportunity (Florea & Mihai, 2015). It is a win–win situation for both employees and the organization to allow the employees for upskilling. Florea and Mihai (2015) have used multivariate regression to address the relationship between the number of training hours and the number of employees undergoing that training and the organizational performance and concluded that the number of training hours has a positive impact on the organizational performance without hiring more employees.

3.3.3.4 Payroll and Compensation

In labor market, salary expectation relies on how productive the labor is and how many years of experience the labor has got. The multiple regression is used to find out salary discrimination based on sex and other variables—namely, the level of education and the number of years in the job. While selecting the dependent variable for salary in the model, choosing the hourly wage is good if the number of work hours is based on the work preference alone and the selection of annual salary is good if there is any discrimination (Rubinfeld, 2011).

3.3.4 Operations

3.3.4.1 Demand Forecasting

When the manufacturing company perceives the demand of the product based on the customer preference, the production and the inventory can be maintained effectively (Guanghui, 2012) along with efficient supply chain collaboration (Carbonneau, Laframboise, & Vahidov, 2008). Not only predicting uncertain demand but also fulfilling that demand with various existing constraints such as proper production schedule and its timely execution, reduction of time between the order and the delivery, and multi-team coordination issues is a challenging task (Guanghui, 2012). Even if the demand is predicted correctly, the company demand will deviate from the actual demand due to the bullwhip effect (Carbonneau et al., 2008). Guanghui (2012) has preferred the support vector regression to forecast the supply chain demand of dairy products and has come to a conclusion that support vector regression gives higher accuracy.

3.3.4.2 Supply Chain Flexibility

Due to climate change and competitiveness, companies have to apply new strategies to handle the supply chain partners because of multiple dimensions involved in the process of the supply chain, namely, the capacity of the company, product delivery, transaction of information with each other, integration with plant and supply chain, and logistics (Jeeva & Guo, 2010). The multi-variate regression analysis is used by Jeeva and Guo (2010) to study the supply chain flexibility, and they concluded that multi-variate regression will be more effective if the number of suppliers is limited.

3.3.4.3 Green Supply Chain Management

Though environmental pollution has been a series issue for a few decades, industrialization is inevitable for country growth paving the way for green supply chain management that drives the organization to enhance organization environmental performance. Various hypotheses, namely, internal and external practices of green supply chain management on organization flexibility, environment pollution, and operational cost are formed and studied using linear regression (Mumtaz, Ali, & Petrillo, 2018).

3.4 Conclusion

The business has been incorporating most of the statistical methods along with state-of-the-technology for many decades. Starting from Microsoft Excel to the advanced technologies uses the regression analysis for identifying the relationship between variables. It is advisable to perform regression in the first place before proceeding to any complex model because regression analysis has been a robust statistical technique for years. Hence, several popular models of regression and its applications are discussed with business cases. Though there are new methods introduced in the artificial intelligence era, regression has its advantages.

References

Aguinis, H. (1995). Statistical power with moderated multiple regression in management research. *Journal of Management, 21*(6), 1141–1158.

Annaert, J., Claes, A. G. P., Ceuster, M. J. K. D., & Zhang, H. (2013). Estimating the spot rate curve using the Nelson-Siegel model: A ridge regression approach. *International Review of Economics and Finance*, 1–15. doi:10.1016/j.iref.2013.01.005

Bager, A., Roman, M., Algelidh, M., & Mohammed, B. (2017). *Addressing multicollinearity in regression models: A ridge regression application*. Munich Personal RePEc Archive, 1–20.

Bansal, A., Kauffman, R. J., & Weitz, R. (1993). Comparing the performance of regression and neural networks as data quality varies: A business value approach. *Journal of Management Information Systems*, 11–32. doi:10.1080/07421222.1993.11517988

Bayaga, A. (2010). Multinomial logistic regression: Usage and application in risk analysis. *Journal of Applied Quantitative Methods*, *5*(2), 288–297.

Bennett, R. (1997). Export marketing and the Interne. *International Marketing Review*, *14*(5), 300–323. doi:10.1108/02651339710184280

Bensic, M., Sarlija, N., & Zekic-Susac, M. (2006). Modeling small business credit scoring by using logistic regression, neural networks, and decision trees. *Intelligent Systems in Accounting, Finance and Management*, *13*(3), 133–150. doi:10.1002/isaf.261

Bergeron, F., Raymond, L., & Rivard, S. (2004). Ideal patterns of strategic alignment and business performance. *Informations and Management*, *41*, 1003–1020.

Bolton, C. (2009). *Logistic regression and its application in credit scoring* [Unpublished Master's thesis]. University of Pretoria, 1–240.

Buteneers, P., Caluwaerts, K., Dambre, J., Verstraeten, D., & Schrauwen, B. (2013). Optimized parameter search for large datasets of the regularization parameter and feature selection for ridge regression. *Neural Processing Letters*, *38*, 403–416.

Carbonneau, R., Laframboise, K., & Vahidov, R. (2008). Application of machine learning techniques for supply chain demand forecasting. *European Journal of Operational Research*, *184*, 1140–1154. doi:10.1016/j.ejor.2006.12.004

Ceh, M., Kilibarda, M., Lisec, A., & Bajat, B. (2018). Estimating the performance of random forest versus multiple regression for predicting prices of the apartments. *International Journal of Geo-Information*, *7*(168), 1–16. doi:10.3390/ijgi7050168

Chattopadhyay, M., Dan, P. K., Majumdar, S., & Chakraborty, P. S. (2012). Application of neural network in market segmentation: A review on recent trends. *Management Science Letters*, *2*(2), 425–438. doi:10.5267/j.msl.2012.01.013

Detienne, K. B., Detienne, D. H., & Joshi, S. A. (2003). Neural networks as statistical tools for business researchers. *Organizational Research Methods*, *6*(2), 236–265. doi:10.1177/1094428103251907

Drucker, H., Burges, C. J. C., Kaufman, L., Smola, A., & Vapnik, V. (1996). Support vector regression machines. *NIPS'96: Proceedings of the 9th International Conference on Neural Information Processing Systems*, *9*, 155–161.

Fasone, V., Kofler, L., & Scuderi, R. (2016). Business performance of airports: Non-aviation revenues and their determinants. *Journal of Air Transport Management*, *53*, 35–45. doi:10.1016/j.jairtraman.2015.12.012

Florea, N. V., & Mihai, D. C. (2015). Improving organization performance through human capital development, using a regression function and Matlab. *Journal of Science and Arts*, *15*(32), 229–238.

Grömping, U. (2009). Variable importance assessment in regression: Linear regression versus random forest. *The American Statistician, 63*(4), 308–319. doi:10.1198/tast.2009.08199

Guanghui, W. (2012). Demand forecasting of supply chain based on support vector regression method. *Procedia Engineering, 29*, 280–284. doi:10.1016/j.proeng.2011.12.707

Hennings, E., & Katchova, A. L. (2005). Business growth strategies of Illinois farms: A quantile regression approach. *2005 Annual Meeting, July 24–27, Providence, RI from American Agricultural Economics Association (New Name 2008: Agricultural and Applied Economics Association)*, 1–29. doi:10.22004/ag.econ.19367

Hopkins, L., & Ferguson, K. E. (2014). Looking forward: The role of multiple regression in family business research. *Journal of Family Business Strategy, 5*, 52–62. doi:10.1016/j.jfbs.2014.01.008

Hosmer Jr, D. W., Lemeshow, S., & Sturdivant, R. X. (2013). *Applied logistic regression* (3rd ed.). Wiley.

Hung, W.-T., Shang, J.-K., & Wang, F.-C. (2010). Pricing determinants in the hotel industry: Quantile regression analysis. *International Journal of Hospitality Management, 29*, 378–384. doi:10.1016/j.ijhm.2009.09.001

Jamaludin, Z. (2009). Perceived organizational justice and its impact to the development of commitments: A regression analysis. *World Journal of Management, 1*(1), 49–61.

Jareño, F., Ferrer, R., & Miroslavova, S. (2016). US stock market sensitivity to interest and inflation rates: A quantile regression approach. *Applied Economics*, 1–13. doi:10.1080/00036846.2015.1122735

Jeeva, A. S., & Guo, W. W. (2010). Supply chain flexibility assessment by multivariate regression and neural networks. *Advances in Neural Network Research and Applications, 67*, 845–852. doi:10.1007/978-3-642-12990-2_98

Kibria, B. M. G., Månsson, K., & Shukur, G. (2011). Performance of some logistic ridge regression estimators. *Computational Economics*, 401–414. doi:10.1007/s10614-011-9275-x

Koenker, R., & Hallock, K. F. (2001). Quantile regression. *Journal of Economic Perspectives, 15*(4), 143–156.

Larasati, A., DeYong, C., & Slevitch, L. (2011). Comparing neural network and ordinal logistic regression to analyze attitude responses. *Service Science, 3*(4), 304–312. doi:10.1287/serv.3.4.304

Liaw, A., & Wiener, M. (2002). Classification and regression by random forest. *R News*, 18–22.

Malthouse, E. C. (1999). Ridge regression and direct marketing scoring models. *Journal of Interactive Marketing, 13*(3), 10–23. doi:10.1002/(SICI)1520-6653(199923)13:43.0.CO;2-3

Mumtaz, U., Ali, Y., & Petrillo, A. (2018). A linear regression approach to evaluate the green supply chain management impact on industrial organizational performance. *Science of the Total Environment, 624*, 162–169. doi:10.1016/j.scitotenv.2017.12.089

Omay, C. (2010). Logistic regression: Concept and application. *Educational Sciences: Theory and Practice, 10*(3), 1397–1407.

Palasca, S. (2012). Emerging markets queries in finance and business statistical evaluations of business cycle phases. *Procedia Economics and Finance, 3*, 119–124. doi:10.1016/S2212-5671(12)00129-3

Pereira, J. M., Basto, M., & Silva, A. F. d. (2015). The logistic lasso and ridge regression in predicting corporate failure. *Procedia Economics and Finance, 39*, 634–641. doi:10.1016/S2212-5671(16)30310-0

Qasim, M., Månsson, K., & Kibria, B. M. G. (2021). On some beta ridge regression estimators: Method, simulation and application. *Journal of Statistical Computation and Simulation*, 1–14. doi:10.1080/00949655.2020.1867549

Rubinfeld, D. L. (2011). Reference guide on multiple regression. In *Reference manual on scientific evidence* (3rd ed., pp. 303–359). Washington, DC: The National Academies Press.

Satheesh, M. K., & Nagaraj, S. (2021). Applications of artificial intelligence on customer experience and service quality of the banking sector. *International Management Review, 17*(1), 9–17.

Sebt, M. V., & Yousefi, H. (2015). Comparing data mining approach and regression method in determining factors affecting the selection of human resources. *Fen Bilimleri Dergisi (CFD), 36*(4), 1846–1859.

Smola, A. J., & Scholkopf, B. (2004). A tutorial on support vector regression. *Statistics and Computing, 14*, 199–222.

Sofie, B., & Hubert, O. (2004). *35 years of studies on business failure: An overview of the classic statistical methodologies and their related problems.* Vlerick Leuven Gent Working Paper Series, 1–70.

Wagner, J. (2004). Export intensity and plant characteristics: What can we learn from quantile regression? *HWWA Discussion Paper, 304*, 1–6.

Wood, E. H. (2006). The internal predictors of business performance in small firms a logistic regression analysis. *Journal of Small Business and Enterprise Development, 13*(3), 441–453. doi:10.1108/14626000610680299

Yu, K., Lu, Z., & Stander, J. (2003). Quantile regression: Applications and current research areas. *The Statistician, 52*, 331–350.

Zuo, Y., Kajikawa, Y., & Mori, J. (2016). Extraction of business relationships in supply chain networks using statistical learning theory. *Heliyon, 2*(6), e00123.

Chapter 4

Chatbots: Their Uses and Impact in the Hospitality Sector

Princy Sera Rajan, Darsana S. Babu, and Sameena M. H.

Contents

4.1 Introduction

In recent times, there is an increasing need for improvement in any industry to keep up with the customer's expectations. Especially in the hospitality industry, as it is a service industry, it deals with the customer service experience, which directly aligns with their satisfaction to be in line and acts as fast as possible; the easiest and immediate solution would be the integration of automated technology using artificial intelligence (AI)-driven chatbots.

AI is the realm of computer science that focuses on the functions of intelligent machines; they mimic human activities and reactions [1]. Some

DOI: 10.4324/9781003206316-4

of the AI-enabled functions include speech recognition, learning, planning, and problem-solving. The conversation is an intelligent device in the AI spectrum, for which it is designed for the specific purpose to meet its user requests. Conversations are understood exactly what the user entries for different details, such as focusing on specific keywords from a conversation between a bot and an individual user [1]. It requires planning and completion of work before the chatbot can move on to its next task and also identifying the file sequence of actions to complete its main tasks there when it comes to complex tasks [2].

Chatbots offer a few key benefits. The first advantage is that the chatbot is available 24/7 to offer App. Another advantage is that the customer waiting time is removed, so customers can access the service any time. For example, in the hospitality context, customers do not have to wait in line to get basic works in a booking room. Chatbots give them the virtual representation of available options and their values and help them make reservations. Another advantage is customization. Interviews work as existing helpers who can keep customer details with the agent continuously so that the operator can provide the client with better information as well as solutions based on current needs and previous discussions [3].

Therefore, they help hotels to customize the customer experience in all stages of the tourist cycle, from the pre-arrival stage to the post-departure phase. This increases product reliability as well as opportunities to visit again and again. Chatbots can also help manage internet traffic and make sure customers get the right kind of information and assistance to improve conversion and reduce stroke. This is very important for businesses that rely heavily on online marketing.

Chatbots can also be used to build customer relationships. For example, a chatbot may be designed to please the customer on his birthday and post promotional offers to visit places. Such actions allow businesses to develop customer experiences while performing marketing activities [4].

4.2 Literature Review

In today's modern generation of service providing, customers expect fast and quality service, which can be provided by using AI. The chatbot is like an application designed to make a conversation with the human.

They are designed to understand the requirement of the person they are having a conversation with and answer with appropriate messages and give recommendations according to their preference. The usage of a chatbot or AI in the field of hospitality is more profitable and cost-effective than having a human agent to address the customers [1]. Chatbots, in the hospitality sector, are involved in hotels' applications for the ease of use with mobile phones; this helps them increase their online presence in social media as well as reach more target customers. They could hold conversations and can communicate with customers from different regions in their language [2].

Even though chatbots have their pitfalls, they have adapted rapidly in the whole hospitality industry from hotel room booking, to enabling services with just a click, to flight bookings [1].

4.3 Proposed Methodology

Considering the methodology for the four research papers mentioned, the following are the inputs, discussions, and the outputs that drive the papers through in and throughout the impact of AI or the use of chatbots in the hospitality industry is shown in Table 4.1.

The percentage of respondents' preference on chatbots from Table 4.2 is shown in Figure 4.1.

The percentage of respondents' feedback on customer satisfaction through chatbot from Table 4.3 is shown in Figure 4.2.

The percentage of respondents' age reacting to AI from Table 4.4 is shown in Figure 4.3.

The percentage of respondents' awareness towards chatbots from Table 4.5 is shown in Figure 4.4.

Concluding Methodology: Having listed the inputs, discussions, and the outputs derived out of all the papers individually, the common objectives of the papers like finding the use of chatbots in the hospitality industry, leading the results to an increased level of customer satisfaction and heavy technology features in the hospitality industry, and taking the baby steps through the chatbots have completed the purpose and the objective of the study through effective input utilization and efficient discussions in Figures 4.1–4.4.

Table 4.1 Different Methodologies

Paper Reference	*Inputs Considered*	*Discussions*	*Resultant Outputs*
Research Paper 1: An explanatory study of customer perceptions of usage of chatbots in the hospitality sector [1].	With about 100 respondents, the paper formed a questionnaire, with all the inputs required for the survey for arriving at the desired outputs. The questionnaire was: 1. Are you aware of the chatbot feature leading hotels have deployed? Have you used it? 2. To what extent do you know about artificial intelligence (AI)? 3. While staying at a hotel, would you like to be served by robots and/or chatbots from check-in to check-out?	With efficient analysis of the data responses received, they differentiated the age, awareness, usage, preferability, and all the required base factors for the study of chatbots in the hospitality sector. The main discussion was about the chatbot injecting any extra customer satisfaction throughout their stay in any hotel. And their outputs are discussed elaborately.	They found the outputs as the following, **Age outputs:** 13% of the respondents were aware of AI; 65% of the respondents had a vague idea of AI; 22% of the respondents were not at all aware of AI. Age group of 25–40 years had a vague idea about AI. **Preferences output:** 10% of the respondents were not open to being served by either; 16% of the respondents were not sure; 69% of the respondents preferred a real person for availing of customer service; 4% preferred a virtual assistant; and 12% preferred both is shown in Table 4.2 **Feedbacks received on customer satisfaction:** 15% said that it depended on the situation; 34% of the respondents felt that technology played an important role in the level of satisfaction with the service; 9% of the respondents thought that a combination of technology and personal touch was required; 29% thought that it Thought that it depended on the service context shown in Table 4.3

Research Paper 2: Smart hotel using intelligent chatbot [5].	The inputs were gathered by the following sources: 1. **NLU**—natural language understanding 2. **NLG**—natural language generation 3. **NLP**—natural language processing 4. Data collection through questionnaire.	The study has used the NLU, NLG, and NLP for driving the inputs into guiding the chatbot through efficient use of AI for guiding the chatbots for use. Customer satisfaction is also a bonus benefit for finding out the uses that chatbots give toward customer satisfaction. The data source of 150 respondents formed the output part of the study. The questionnaire is also supported for the sake of finding customer satisfaction. The studies also provide the art of making its services more unique, efficient, and effective for the users and the readers.	The plotting of the user inputs and the recognition of various features of the language is done through the NLU, while the NLG is used for text planning (for related and relevant chatbot replies), sentence planning (grammatical formation and the replacing with the right choice of words, phrases, and the fundamental attitude and behavior in the conversation, i.e, on subtle grounds), text realizations (structuring out the sentence on the basis of the aforementioned factors), and NLP is used for computerized formats (by gathering the knowledge on how the humans act on various situations and answer on various questions through tools, codes, program, techniques that are designed in the chatbots based on the aforementioned document analysis). Results of the questionnaire also drive the output results as follows: **The 150 sampled questionnaire showed results of the following consumer feedback results:** 40% of the respondents were already satisfied with the provided privacy, safety, and the security of the hotel they checked in. While 20% were not satisfied, the reason mentioned differed. They specified the location of the hotel is the reason for their decision, while 40% felt that their consumer experience can adapt to the new grown technologies and other facilities provided to the customers who checked in the latest, through proper awareness about the hospitality industry and the awareness about the use of AI in the same is discussed in Tables 4.4 and 4.5

(Continued)

Table 4.1 (Continued)

Paper Reference	*Inputs Considered*	*Discussions*	*Resultant Outputs*
Research Paper 3: Chatbots—an organization's friend or foe? [2].	The paper has done its research about the use of chatbots in the hospitality industry. The authors also discuss the rapidity it causes in various sectors and has also provided ample factors for increasing its impact in theindustry. The inputs have also been improvised through efficient resource utilization.	Through qualitative data being sieged and provided, the inputs are incurred in the effects and are also categorized into benefits and pain points. With the use of chatbots, the paper shows that the customer experience has improved, thus the hotels in the hospitality sector are developing themselves in the technology or the AI aspects.	Through pre-programmed conversations and suggesting places that the user might want to explore, the chatbot provides all the details by understanding and responding. By customizing all the languages, chatbots are the surprising advantages for all the hotels using AI as their tool of development. Thus, AI is capturing its steps into the hospitality industry through chatbots is discussed in Tables 4.4 and 4.5 Chatbots can manage numerous channels and their customers at the same time. The Internet, being used for numerous reasons, is now efficient through chatbots through travel agency apps like MakeMyTrip, Golbibo. Thus, through a signature move, the hotels are just a click away from the customer. Still, chatbots come with a package that includes limitations like privacy, less intuitiveness, less awareness, and the chances of consumers building an uncomfortable feeling.
Research Paper 4: Chatbot adoption in tourism services: A conceptual exploration [3].	1. Understand the factors influencing the firm-level adoption of chatbots by tourism and hospitality industries. 2. Examining the role of chatbots in various areas of the tourism and hospitality industry.	We adopted two core organizational theories: institutional theory and organizational learning theory.	Based on isomorphic pressures, learning capabilities, learning competencies, and barriers are the control variables and adoption intention is the dependent variable. Almost 85% of all interactions will be carried out through chatbots in future.

Note: The methodology of the Research Paper [1] (an explanatory study of customer perceptions of usage of chatbots in the hospitality sector) has also dipped an information of customer awareness of within the hospitality industry. *The summarized outputs of the research paper are figuratively explained below.*

Table 4.2 Percentage of Respondents' Preference on Chatbots

1	10%	The respondents were not open to be served by either
2	16%	The respondents were not sure
3	69%	The respondents preferred a real person for availing customer service
4	4%	Preferred a virtual assistant
5	12%	Preferred both

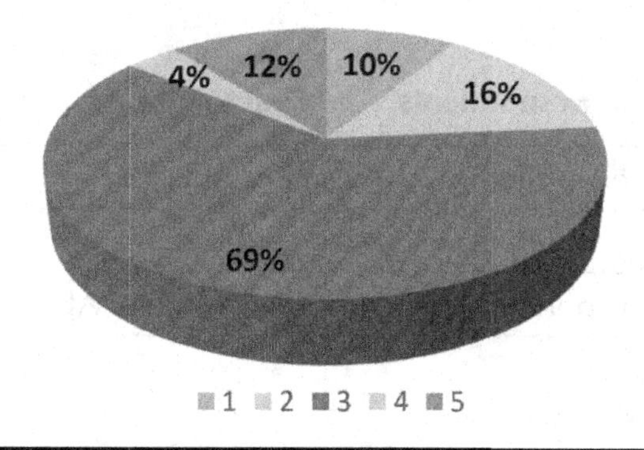

Figure 4.1 Percentage of respondents' preference on chatbots.

Table 4.3 Percentage of Respondents' Feedback on Customer Satisfaction through Chatbot/AI

1	15%	That it depended on the situation.
2	34%	The respondents felt that technology played an important role in the level of satisfaction of the service.
3	9%	The respondents thought that a combination of technology and personal touch was required
4	29%	Thought that it depended on the service context.
5	28%	Thought that technology played no role in the level of satisfaction.

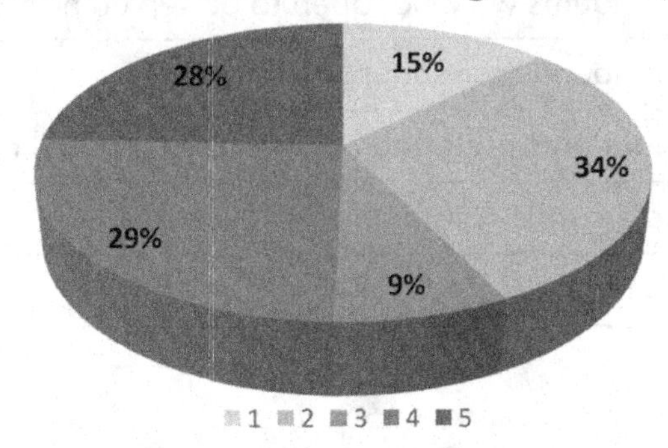

Figure 4.2 Percentage of respondents' feedback on customer satisfaction through chatbot/AI.

Table 4.4 Percentage of Respondents' Age Reacting to AI

1	13%	The respondents were aware of AI.
2	65%	The respondents had a vague idea of AI.
3	22%	The respondents were not at all aware of AI.

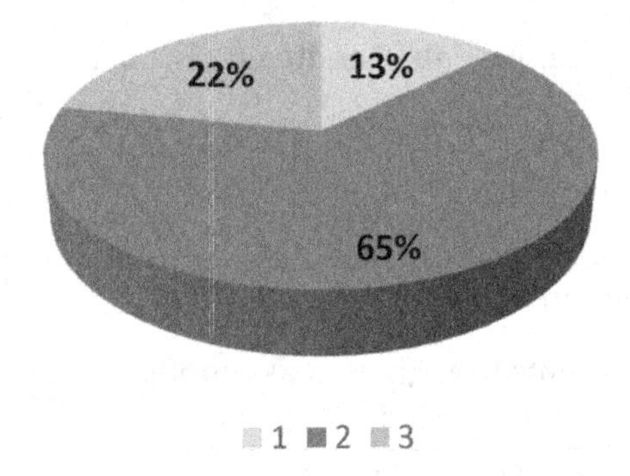

Figure 4.3 Percentage of respondents' age reacting to AI.

Table 4.5 Percentage of Respondents Awareness Toward Chatbots

1	6%	The respondents knew about chatbots and had used them.
2	60%	The respondents had not even heard of chatbots.
3	4%	The respondents had heard of them but were not sure what they were.
4	30%	The respondents knew about chatbots but had never used them.

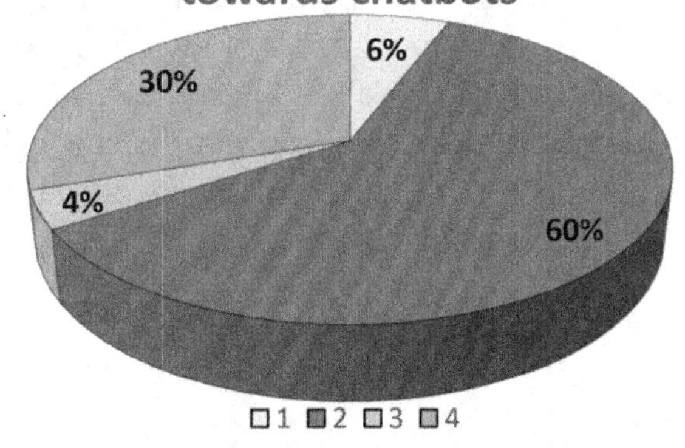

Figure 4.4 Percentage of respondents' awareness toward chatbots.

4.4 Limitations of Using Chatbots

Using chatbots can have many advantages. Collectively, there are many limitations described below.

- As chatbots are being designed using neural networks and machine learning (ML), it will be impossible in some cases to give a response to critical customer queries.
- Decision-making skills for chatbots are very low.
- These chatbots are very least ranked in customer retention.

- In this crucial and important sector like hospitality, it is very important to handle a customer with convincing comments and answers.
- Chatbots are very effective in reducing manpower such as helpdesk and reception.
- So, chatbots are usually programmed to answer for a certain set of questions using ML but if a customer raises queries for any other question rather than trained questions, it may give the same answer that does not match.
- Therefore, chatbots are lagging in this decision and critical thinking skills.

This book speaks about the usage of chatbots in the hospitality sector. These assumptions are being made by taking the tourist sector as the key factor. We analyzed how customers think of aware chatbots. We took respondents' opinions on chatbots like how often they use chatbots; how they are about these automated answering algorithms; and which ages more frequently use these chatbots. This chapter finally speaks about customer satisfaction and its advantages in mere future.

4.5 Conclusion

Technology and innovation are in a fast-running phase in the present world. Chatbots can be very advantageous in the hospitality sector for giving suggestions within the hand without asking anyone in person. This reduces the time and effort of an individual to make decisions at an instant and proceed. These chatbots give suggestions based on ML and neural network algorithms. These applications can give real-time solutions from the previously trained data sets and feedback. This helps in giving more accurate suggestions and predictions than humans do. When we speak for the tourism sector, it is very important for finding true visiting spots and locations. In this application, chatbots can be very useful in answering the user inputs. Although there are limitations like decision-making skills and critical thinking skills, they can be corrected by giving updates to these chatbot applications or websites.

4.6 Future Scope

As this analysis speaks about only four researches, it can be made better by adding more research articles into consideration. Further work can be done in analyzing how to increase decision-making and risk-taking skills in AI chatbots. However, analyzing this particular topic requires more time and effort.

References

[1] Mihir Dash & Suprabha Bakshi, An Exploratory Study of Customer Perceptions of Usage of Chatbots in the Hospitality Industry, *International Journal on Customer Relations*, 7 (2), 2019, 27–33.

[2] Emma Carter & Charlotte Knol, Chatbots—an organisation's friend or foe?, *Research in Hospitality Management*, 9 (2), 2019, Stenden Hotel Management School, NHL Stenden University of Applied Sciences, 11.

[3] Dandison Ukpabi, Bilal Aslam & Heikki Karjaluoto, Chatbot Adoption in Tourism Services: A Conceptual Exploration. In Stanislav Ivanov & Craig Webster (Eds.), *Robots, Artificial Intelligence, and Service Automation in Travel, Tourism and Hospitality*. Emerald Publishing Limited, 2019, 105–121. DOI: 10.1108/978-1-78756-687-320191006.

[4] Santosh Bisoi, Mou Roy & Dr. Ansuman Samal, Impact of Artificial Intelligence in the Hospitality Industry, *International Journal of Advanced Science and Technology*, 29 (5), 2020, 4265–4276.

[5] Shubham Parmar, Megha Meshram, Parth Parmar, Meet Patel & Payal Desai, Smart Hotel Using Intelligent Chatbot: A Review, *International Journal of Scientific Research in Computer Science, Engineering and Information Technology (IJSRCSEIT)*, 5 (2), March–April 2019, 823–829, ISSN: 2456–3307. https://doi.org/10.32628/CSEIT11952246. Journal URL: http://ijsrcseit.com/CSEIT11952246

Chapter 5

Traversing through the Use of Robotics in the Medical Industry: Outlining Emerging Trends and Perspectives for Future Growth

Gaurav Nagpal, Kshitiz Sinha, Himanshu Seth, and Namita Ruparel

Contents

DOI: 10.4324/9781003206316-5

5.1 Introduction

Humans and artificial intelligence have joined forces in numerous fields with machines [1]. Robots have become the reality of the contemporary world, enabling humans to become more effective and connected [2]. There are many verticals in which robots are being used, having specific purpose and utility in each of them, and designed accordingly to assist the humans wherever possible with higher accuracy and precision. The technology has been increasing exponentially since the past decade. Every industry is opting for inducing its average workflow with robotic technology as it has proven to be beneficial not only in terms of profit but also in terms of reduced human error and increased dexterity and accuracy. Also, this is the age of AI with increasing automation and decreasing human intervention in the processes. The amount of capital being invested in AI by the companies, as well as the number of people studying AI, has witnessed an increasing trend.

Mentoring and on-the-job training in the field of medical surgeries can also be done with the help of robots the way it is being resorted to in the other industries [3]. Any job that has an outcome which can be pre-stated or guessed should ideally be done by machines so that the humans can have the bandwidth to perform the open-ended jobs [4]. A risk-oriented framework has been proposed by Dhar (2015) to evaluate which decisions should be made by robots and which ones by humans [5].

The healthcare sector is one of the fields in which advancement is far from saturation. New directions of research are being discovered every year that calls for endless innovations. In this chapter, the authors describe the prospects of robotic technology being used in the healthcare sector, specifically focusing on surgical robots, the development of which is in a nascent phase. The need for robotic surgery will eventually prevail because, with the increasing population and medical awareness, the number of surgeries is also expected to increase, leading to more demand for surgeons. Here, robotic surgery can be a solution amidst the lower per-capita availability of skilled medicos, as the operation time and burden on the surgeon will both be reduced with robotic surgeries. Not only this, the recovery after a robotic operation is proven to be faster than open surgery, leading to shorter hospital stays which means more rooms for patients and lesser hospitalization costs.

Robotics surgery is defined as the involvement of robotically assisted surgical (RAS) devices that enable the surgeon to use computer and software technology to control and move surgical instruments through one or more tiny incisions in the patient's body (minimally invasive) for a variety

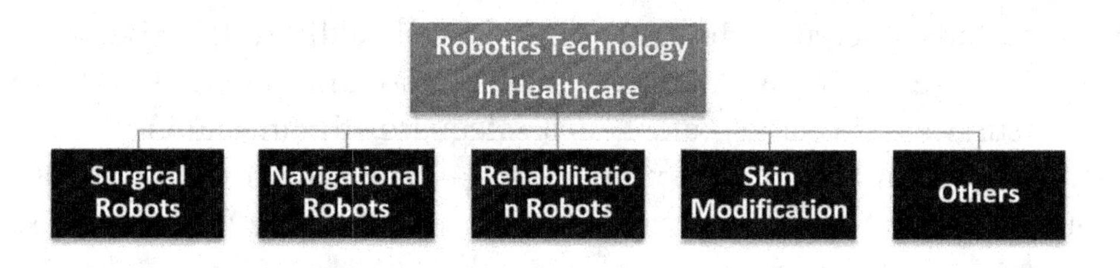

Figure 5.1 Classification of robots in the healthcare industry.

***Source:* Self-composed by the authors.**

of surgical procedures. RAS devices generally have three main components: a console, the bedside cart, and a separate cart containing the supporting hardware and software [6]. The healthcare industry is undergoing automation with the rising number of robotic technologies. Various types of robots are being used at present, each one having its unique features and applications in surgery. The most basic classification is shown in Figure 5.1.

Each classification in this figure has a distinct approach to surgeries. Surgical robots are used for performing normal operations on humans such as general, laparoscopic, brain, spine, ocular, and cardiovascular which involve cutting or probing specific parts (electrodes and knee implants). Navigational robots are used only for diagnosis and scanning of various body parts for any pharmacological action, used in endoscopy. Three-dimensional (3-D) and high-definition (HD) vision can be obtained by the robotic arm, used to determine if the operation is required for the individual or not. Rehabilitation robots are robots that help in improving the body movements of people who were paralyzed or had impaired body movements. Robots can provide automatic body movements which can aid the elderly and disabled. Skin modification robots are also available which are used to obtain wrinkle-free and rejuvenated skin. The category "Others" comprises hygiene maintaining robots that are not directly related to human health but are a necessity in hospitals for maintaining sanitation, collection of records through face recognition, dispensing of medicines, and in many other operations. Each type of surgery requires different techniques to operate with and has a different set of patents. Various technologies have been covered with innovations that are being developed in this industry. Following the trend, major companies have started to invest in this sector.

Numerous procedures need to be followed even after a robot has been developed and tested, before it is available to the population. As stated earlier, surgeries of different parts of the body require a different technology to operate with and have a different set of patents. Newly developed innovations and discrete technologies in robotic healthcare have been covered in this research study. An aspect that is important to understand while studying surgical robotics is patent, as it safeguards the invention of the company and allows the inventor to enjoy market exclusivity throughout its operational time and also is a major criterion when distinguishing robots.

5.2 Evolution of Robotic Surgery

The first-ever documented robot-assisted surgery (RAS) was carried out in 1985, with a device known as PUMA 560 (Programmable Universal Machine for Assembly or Programmable Universal Manipulation Arm). It was a single-standard robotic arm, developed by Unimation, and was used to guide a needle to the brain for biopsy, a procedure previously subjected to error due to hand tremors. It was made possible by implementing computer tomography (CT) for guidance. In 1998, PROBOT was developed by Imperial College, London, and was used to perform transurethral prostate surgery as the procedure required frequent repetitive cutting motions.

The next robot in line was ROBODOC, 1992, developed by Intuitive Surgical Systems and IBM; it was used for creating a cavity in the femur for hip replacement. The next in development was Automated Endoscopic System for Optimal Positioning (AESOP) system, in 1990, by Computer Motion, which was the first system approved by the US Food and Drug Administration (FDA) and was used for endoscopic surgical procedures. ZEUS Robotic Surgical System was developed by the same company, designed to assist in surgery. DaVinci Surgical System, the World's most used robotic surgical, was approved by the United States Food and Drug Administration (US FDA) in 2000. Then, Intuitive Surgical bought Computer Motion, thus discontinuing usage of the ZEUS Robotic Surgical System. Till present, DaVinci has dominated the market of robotic surgery. Many other robots have come into existence in the past decade and have achieved a successful hold in the market in the field of their usage.

5.3 Benefits of RAS

RAS has various advantages over traditional open surgery. The robots act as extended and ergonomic hands of the surgeon that have better capabilities and also cause less cutting and operation. The doctor can sit on a console or use a hand-held device that can make the job much easier, hence enabling him to perform a larger number of operations within the prescribed time, with less stress and workload. RAS decreases the hospital stay time since it avoids basic cutting (minimally invasive surgery); therefore, the patient can get back to his physical strength in a matter of few days, in contrast to a couple of weeks if performed traditionally. Since the recovery time is less, the patient can get back to his normal life without having much to worry about. RAS results in a lesser amount of blood loss and suturing. This procedure enables better visualization of the internal structure of the body, through various interfaces in HD and 3D, which can act as eyes of the surgeon. A few years ago, surgeons had to move the laparoscopic camera by hand, and it only offered a vision in a single direction, but there have been developments of a robotic camera that offer a wider field of vision and thinner arms to provide easier penetration. Gesture-controlled cameras are also present that enable a surgeon to control the camera according to his comfort. Some of the software can take various MRI scans and combine them to provide a single, refined image of our brain. Simulation software has also been developed to provide lifelike operating experiences to the trainee who is learning the use of robots. Knee and hip replacement robots offer movement in 7 degrees of freedom and simulation of hand movements that provides efficient placements of implants. The software also determines a pre-operating plan that is to be followed by the robot, and if by any chance a deviation occurs, it can be overridden manually.

5.4 Current Market Scenario

Diving into surgical robots, they can roughly be categorized into broad classes, as shown in Figure 5.2.

Single Port Access (SPA) is known as the process in which a small incision (2′–4′) is made on the body instead of a whole cut, thereby offering several advantages:

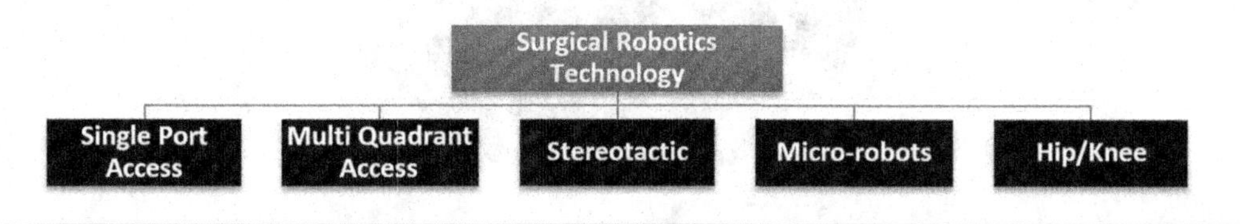

Figure 5.2 Classification of technology used in surgical robots.

- Less amount of blood loss and lesser quantity of blood required during the procedure
- Less post-operative stay of the patient in bed (2–3 days) which is much higher in general open surgery (2 weeks to 1 month)
- Less stress on the surgeon therefore can perform a higher number of operations
- Lesser room in making mistakes as human error is eliminated
- Minimal scarring as the operation is conducted using minimally invasive surgery
- The amount of pain and discomfort is significantly less
- Better visualization of the internal body
- The ergonomic design of the robotic arm end.

SPA is used almost in all of the surgeries available to a human being—general, spine and brain, cardiovascular, vascular catheterization, gynecology, ocular, neurology, and urology. Figure 5.3 shows that 36% of the SPA robots are currently in the pipeline.

Various companies have their commercialized robots and some of which are still in the pipeline. Some of the most commonly used robots in the industry are DaVinci (Intuitive Surgical), MAZOR X (MAZOR), Senhance (TransEnterix), and Excelsius GPS (Globus Medical).

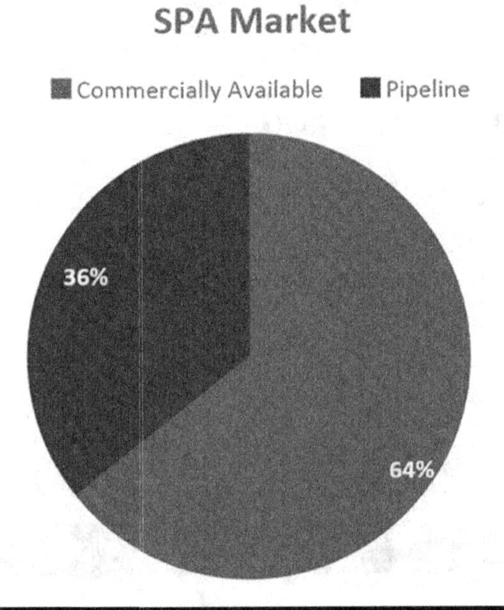

Figure 5.3 The breakup of Single Port Access market of surgical robots.

The number of robots currently in the market is larger in number than those that are still in development. One of the most awaited and currently in development is Verb Surgical, which is being headed by Alphabet (part of Google), and its aim is to "democratize surgery." It will have AI augmentation, which is supposed to change the field of surgery and is not launched completely yet.

Some technologies are required in all of the surgeries and are a must-have for all the robots:

- Full wrist-like rotation of the robotic arms end to provide ergonomic access
- Variable grip graspers so that the arm can be used for various functions such as cutting, suturing, grasping
- Planning out preoperative 3D plans which are to be followed by the robot thoroughly
- Mapping out a virtual boundary so that the robot doesn't go around and harm healthy tissues and organs
- Providing real-time imaging so that the surgeon can see where and how it is operating through an endoscopic camera
- Providing haptic feedback in various degrees to create a sensation of touch when operated by the robotic arm, which can help the surgeon in identifying different tissues without physically feeling the tissue by his/her hand.

These robotics technologies are still present only in well-developed countries, while each company is trying to make its product available for the worldwide population. Looking at the following graph we can see the growth in the usage of robots in performing total knee arthroplasty in the United States from 2005 to 2014. There were a total of 60,60,901 cases reported from 2005 to 2014 in which 2,73,922 were performed using computer navigation and 24,084 were completed using robotic assistance. A graphical representation of the data is shown in Figure 5.4 [7].

According to the findings of a study by BCG, as of January 2016, the global medical robotics industry clocked an annual revenue of $7.47 billion, and this was destined to rise at a robust compounded annual rate of 15.4% over the next five years. Considering this, the global sales of medical robots are expected to grow by $8.90 billion during 2022–2026 [8].

Even in the Middle East countries such as Saudi Arabia, there has been considerable growth in robotic surgeries. The total number of RAS in the Middle East is still low compared to Europe and the United States [9]. To

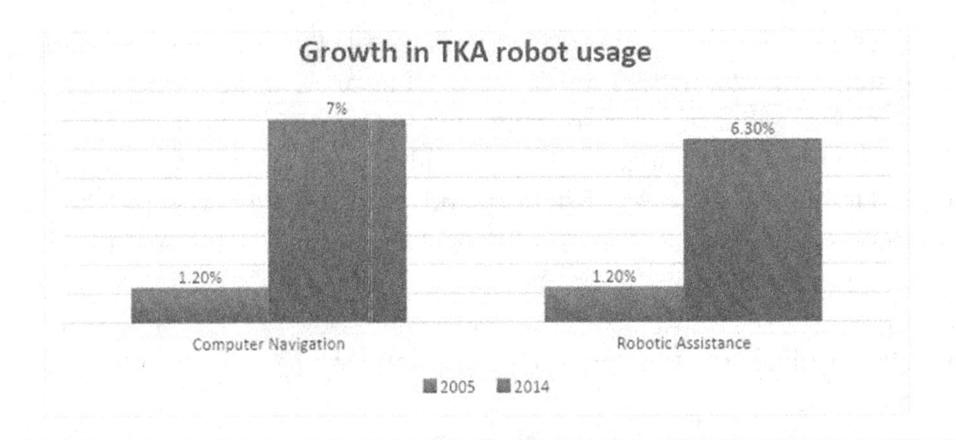

Figure 5.4 The growth in robots usage from 2005 to 2014.

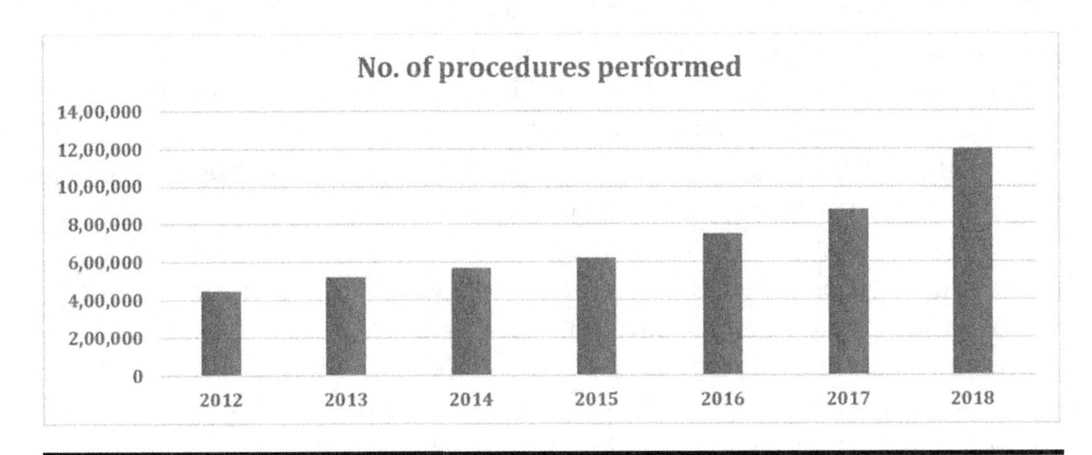

Figure 5.5 The trend of the number of procedures performed globally in 2017.

date, there are over 44 hospitals with DaVinci robots installed in them and ready to use for the general population [10].

Over the years, Intuitive Surgical has seen an exponential rise in usage of their robots for various procedures such as urology and gynecology. Figure 5.5 depicts the approximate number of RAS procedures that have taken place all over the world.

Similarly, New Zealand was introduced to robotic surgery in 2017 by Intuitive Surgical at Southern Cross [11]. Since then, the number of operations being held there is increasing significantly every year. Most DaVinci robots are still in the United States, with the number being over 2,500, followed by China and Australia.

South Korea is also not behind the other nations in using the robot to conduct surgeries. A study shows the usage of DaVinci at a single institute from 2005 to 2013 had steadily risen and seen the trend it will continue to

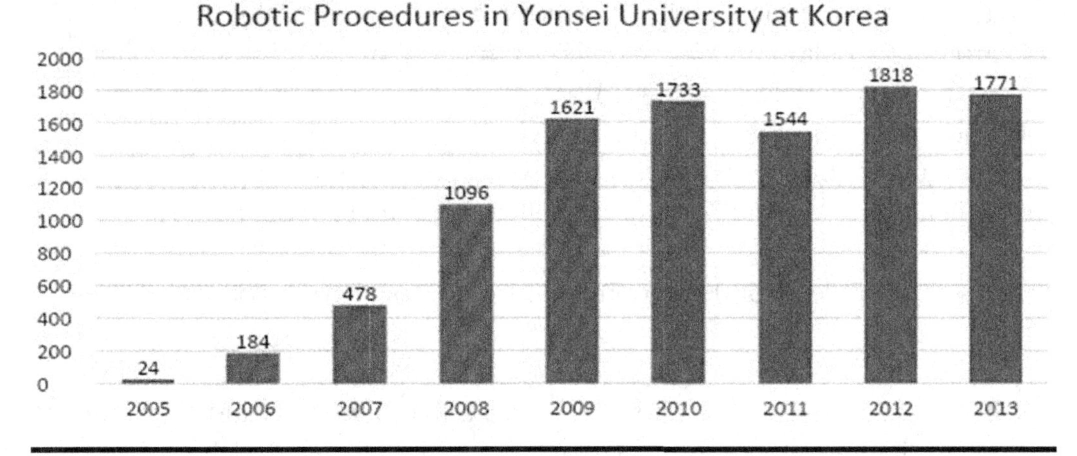

Figure 5.6 The growth in the robotic procedures performed in Yonsei University in Korea.

grow in future also. During this period, robotic system surgery was used for a total of 10,267 cases. Statistical analysis showed that 15 surgeons performed 93.47% of all 10,000 surgeries. One of the surgeons performed more than 2,500 surgeries, and two performed more than 1,400 each. Figure 5.6 shows the distribution of the mentioned surgeries over the years [12].

Figures 5.5 and 5.6 provide us with a general idea about how robotic surgeries are getting a grip on the industry.

This sector, which was dominated by a single company for the past few decades, may witness fragmentation in the near future due to the development of many surgical robots worldwide. Intuitive Surgical with its DaVinci series of robots has dominated the global market and is a pioneer in this industry. It continues to be a prominent developer and seller of the most widely used robots in various surgeries and the most recommended ones by various surgeons. To cite an example, MAKO is the robot that is used extensively in total/partial knee/hip replacement surgeries. The market is getting disrupted by many companies that are developing robots of their own, of which one of the most awaited is Verb Surgical developed by Alphabet (part of Google).

5.5 Future Growth Estimates

Following are the estimates that the author has found through secondary research or has quantified according to previous data collected from various statistics present:

- The worldwide market size of robotic devices is currently evaluated at around $5 billion and is expected to multiply by five times the amount in the next couple of years [13].
- The market in India for medical robotics is going to reach as high as Rs. 2,600 crore by the end of 2024, as stated by the company Vattikuti, which plans to install a total of over 100 DaVinci robots by 2020.

And also train at least 500 surgeons on how to operate the robot to its full potential [14].

- The system revenue of Intuitive Surgical, manufacturer, and distributor of DaVinci robot would grow close to $1.5 billion by 2021 from $1.13 billion in 2018. It is presently having over 4,900 functional units all over the world. Assuming constant growth, the number is expected to grow to 6,000 by 2022 [15].
- A greater number of devices can be found in the United States, followed by Europe and Asia. Every one in four hospitals in the United States has a surgical robot. The manufacturers are planning to target highly populous countries like China and India for increasing their usage.
- The official notification issued by the Chinese Government had expressed the requirement of 197 endoscopic surgical instrument control systems, or surgical robots, by the end of 2020. That number includes an additional 154 new systems of Intuitive Surgical [16].
- Intuitive Surgical is also aiming to set up their base in the military field also, as using 5G technology complex surgeries can take place without the requirement of a surgeon to be on field duty, but just controlling the robot from his/her convenient place.

5.5.1 Reasons for the Lower Than Required Presence of Surgical Robots

- The main reason for the under-penetration of robots in healthcare, which has been stated by the experienced professionals and experts in this field regarding better exposure to robotic surgery, is the lack of clinical evidence.
- Another reason is the reluctance of the ecosystem to change and adapt with time. There is a lot of potential in this market, which people fail to recognize for the fear of loss of jobs or fear of upskilling themselves. We need some organizations that deal with gathering this evidence and

publishing it, to make people who are unaware aware of the various benefits and advantages of RAS over traditional open surgery.

- Advertisement is another point to work upon. The medicos or the customers must get to know about any product that comes to the market. For this, the RAS device manufacturers need to promote the robotic surgeries to the doctors, who, in turn, need to promote them to the patients.
- The medicos have a lack of skill that is required for robotic surgeries. If this skill can be imparted to the surgeons, the robotic surgery can have the potential for inclusive healthcare, as the robots can be updated with the latest software updates equally and the surgeon can operate on the patient sitting at his/her desired location (Telesurgery).
- The robotic parts are costly to make, and they require replacement and maintenance at a regular frequency, making it more costly unless the sufficient scale is achieved.
- There is a loss of human touch of the doctor when the patients get attended to by a robot, which gives RAS a lower edge over the manual surgeries.

5.6 Limitations of Robotics

Till now, we have seen the benefits and the bright side of the application of robotics in healthcare. However, like every other innovation, it has its disadvantages or hurdles that need to be surpassed to increase its usage.

- The most important of them is the learning curve required for the operation of these robots. One just can't come and start using them; a surgeon undergoes various months of training to learn the abilities and working of robotic assistance and to get the full potential out of it. The patient stays at the hospital may be lessened, but the time required for trainees to become proficient in the use of robots is much higher.
- The second advantage is the high cost over a small scale. Any technology cannot become successful until and unless it is being used by the common population, and looking at current numbers, it won't catch up with the people by costing nearly $2,000 per operation. Further, the annual maintenance of the robot is also required, which adds up to several hundred thousand dollars for the hospital. The price of each machine goes up to "$1 million to $2.3 million, plus up to $170,000 per year in maintenance, according to Intuitive Surgical" (Torres, 2011) [17].

- Thirdly, some robots require the patient to be in a certain position for the robot to operate on them. If the patient lies in improper positions during the whole operation, there may be cases in which the patient cannot feel his/her legs or hands, called partial paralysis or positioning injury.
- Fourthly, the usage of telesurgery requires a good data connection. A fault that may occur in the absence of a strong internet connection may be beyond the capabilities of the surgical team. Hence, all the risks must be kept in mind, and countermeasures should be ready for each failure mode that can occur.
- Fifthly, the issue with latency, which is the time required by the robot to execute the command given by the robot, can create a problem. It can create a problem for a surgeon to respond if any complications occur.
- Sixthly, the RAS devices have relatively larger footprints on the space and relatively more cumbersome robotic arms. This gives them a hands-down in the crowded operating rooms, where it may be hard for both the surgical team and the robot to fit into the operating room.
- The seventh disadvantage is the psychological fear of the population. People are afraid to go under the knife and being operated on by a robot. They feel more secure, if a human is performing the operation, and psychological behavior plays a very important in the well-being of a patient's health post-operation.

5.7 Future Directions for Increasing the Acceptance of Robotic Surgeries

There are still various voids that need to be filled and limitations that need to be overcome through the innovations. Some of the future directions are listed below:

- The operating system of RAS can be made more user-friendly and well defined so that it can be learnt much faster, and augmentation with AI to suggest further operations is also possible, which would save the time of both patient and surgeon.
- The robotic arms need to have ergonomic features so to be able to operate through any angle without the hassle of the positioning of the patient. This will prove to be beneficial for both the surgeon and the patient.

- The issue of latency can be solved by involving the technology of 5G, which will also help in telesurgery as it involves the transfer of heavy amounts of data in real time.
- The development of better laws and coverage by insurance in case of mishaps during robotic surgery can help the patient in terms of money and future operations.
- Currently, there are two main agencies that facilitate the verification and commercialization of surgical robots, USFDA (followed in the United States and some Asian countries) and EMA (followed in Europe). If there can be a single, unified agency for this job, the distribution and commercialization can be much easier, and we can do away with the need of applying in every administration before marketing.
- US Army and NASA scientists are taking a heavy interest in the development of telesurgery to be able to attend to their soldiers and astronauts in space so that the requirement of the physical presence of surgeons is eliminated.
- Software being used in robots needs to be upgraded constantly to make it user-friendly and provide a better user interface and lesser lag time in between commands and execution.
- Various levels of haptic feedback can be developed to enable robotic surgery in delicate operations such as optical surgeries.
- The development of a computer-based intelligent tutoring system for the training of cognitive and procedural skills that are needed to complete basic robotic suturing can be useful for novice surgeons [18, 19].
- Insurance companies need to be encouraged to also include robotic surgery in the healthcare cover, which can also help in inducing the sense of assurance that robotic surgery is safe and prominent future of healthcare.

5.8 Conclusion

This chapter explains the current landscape of the robotic surgery industry. The investment required in the development and operation is large, and hence institutions that choose to acquire the RAS devices should plan out a cost-effective method for their utilization. Currently, the geographical market that has been captured by the medical robots is not diverse and can be seen in heavy usage only in few top developed countries of the world, the United States, of course, being at the top, followed by SEA, Europe, and Australia.

Although robotic surgery is currently being utilized in developed countries mostly, in the next decade, it will achieve a worldwide presence. It will be replacing the extra physicians who need to be present in the operation theatre, with the presence of just the surgeon and the patient, thereby lessening the workforce required. The benefits of minimally invasive surgery are far greater than the risks it has; and since it can't be performed by human hand, robotic surgery will become a better option in the coming future. It will bring significant changes to healthcare.

As conveyed in this study, the authors feel upbeat on the future of this industry and move on to suggest that robots are the future of surgery because robotic surgery has various advantages over general open surgery, some of them being lesser post-operative trauma, lesser operative time, precise robotics movements, a lower amount of blood required for operation, better and faster decision-making on the diagnosis and the course of action, better and real-time documentation of the medical records of the patients, a wider range of complicated surgeries that can be performed by robots, and mass customization of the surgeries using AI and ML.

This study claims that robotics is the next big innovation in the healthcare sector as the market has been continuously growing for the last five years and will to grow exponentially in the coming years. The usage of these robots is also growing as people are becoming aware of their benefits and shunning the fear of being operated by a robot. Development in this sector is only waiting to be explored as we can develop a new development/news articles/published papers every other day, and the companies are more ready than ever earlier to inculcate them and also bank upon them.

The chapter also puts forward how robotics can help in medical inclusion, particularly in the nations that have very low per-capita availability of skilled medical doctors. The authors express their belief that a time will come when robotic surgeries will be much more economical than manual surgeries. Finally, the study also cautions about the flip side of using robotics in healthcare. One of the areas touched upon includes the loss of the personal touch in the treatment of a patient by a robot.

References

[1] Wilson, H. J., & Daugherty, P. R. (2018). Collaborative intelligence: Humans and AI are joining forces. *Harvard Business Review*, July–Aug.

[2] Joel, M. (2012). Don't be afraid of the robots. *Harvard Business Review*, Dec 18.
[3] Beane, M. (2019). Learning to work with intelligent machines. *Harvard Business Review*, Sep–Oct.
[4] Nedelescu, L. (2015). We should want robots to take some jobs. *Harvard Business Review*, June 5.
[5] Dhar, V. (2016). When to trust robots with decisions and when not to. *Harvard Business Review*, May 17.
[6] US Food and Drug Administration Website. www.fda.gov/medical-devices/surgerydevices/computer-assisted-surgical-systems
[7] Antonios, J. K., Korber, S., Sivasundaram, L., Mayfield, C., Kang, H. P., Oakes, D. A., & Heckmann, N. D. (2019). Trends in computer navigation and robotic assistance for total knee arthroplasty in the United States: An analysis of patient and hospital factors. *Arthroplasty Today*, *5*(1), 88–95
[8] Chengdu Tianfu Software Park. (2018). *Borns: Breakthrough and prospect of AI-powered minimally invasive surgery*. www.prnewswire.com/news-releases/borns-breakthroughand-prospect-of-ai-powered-minimally-invasive-surgery-300675230.html
[9] Rabah, D. M., & Al-Abdin, O. Z. (2012). The development of robotic surgery in the Middle East. *Arab Journal of Urology*, *10*(1), 10–16. Doi:10.1016/j.aju.2011.12.001
[10] Azhar, R. A., Elkoushy, M. A., & Aldousari, S. (2019). Robot-assisted urological surgery in the Middle East: Where are we and how far can we go? *Arab Journal of Urology*, *17*(2), 106–113. www.ncbi.nlm.nih.gov/pmc/articles/PMC6600062/
[11] Ford, E. (2016). *Surgeons perform first robotic surgery in New Zealand to treat oral cancer*. www.stuff.co.nz/national/health/86913698/surgeons-perform-first-robotic-surgery-in-newzealand-to-treat-oral-cancer
[12] Koh, D. H., Jang, W. S., Park, J. W., Ham, W. S., Han, W. K., Rha, K. H., & Choi, Y. D. (2018). Efficacy and safety of robotic procedures performed using the da Vinci Robotic Surgical System at a Single Institute in Korea: Experience with 10000 cases. *Yonsei Medical Journal*, *59*(8), 975–981. https://doi.org/10.3349/ymj.2018.59.8.975
[13] Jinoy Jose, P. (2018). *Da-Vinci's 5 million surgeries*. A blog article. www.thehindubusinessline.com/opinion/columns/the-cheat-sheet/da-vincis-5-million-surgeries/article24873712.ece
[14] Trefis Team. (2019). *How much can intuitive surgicals system revenue grow over the next three years*. A blog article. www.forbes.com/sites/greatspeculations/2019/07/09/how-much-canintuitive-surgicals-system-revenue-grow-over-the-next-three-years/#67d975bf4418
[15] Hartford, J. (2015). Intuitive surgical makes the case for robotic surgery. *Medical Device and Diagnostic Industry*, *37*(7).
[16] Dansford, F. (2018). *Intuitive surgical shares rise on expanded surgical robot quota in China*. A blog article on Mass Device. www.massdevice.com/intuitive-surgical-shares-rise-onexpanded-surgical-robot-quota-in-china/

[17] Toress. (2011). *Political implications and robotic surgery.* A blog article. www.*bartleby.com/essay/Political-Implications-And-Robotic-Surgery-PKLNH6FMZRPA*
[18] Bertelsen, A., Melo, J., Sánchez, E., & Borro, D. (2013). A review of surgical robots for spinal interventions. *International Journal of Medical Robotics*, 9(4), 407–422. Doi:10.1002/rcs.1469
[19] Julian, D., & Smith, R. (2019). Developing an intelligent tutoring system for robotic-assisted surgery instruction. *International Journal of Medical Robotics and Computer Assisted Surgery.* https://onlinelibrary.wiley.com/doi/abs/10.1002/rcs.2037

Chapter 6

Integration of AI in Insurance and Healthcare: What Does It Mean?

A. Kannan, B. Justus Rabi, and M. Anand

Contents

DOI: 10.4324/9781003206316-6

6.1 Introduction

Insurance often historically had a poor level of consumer involvement. Insurers are the industry in which consumers engage the least, according to the study. As insurance and healthcare is not completely digitized, the healthcare and insurance providers did not get any opportunity to actually understand what the customer needs and offer tailor-made solutions [1].

6.1.1 The Industry's Major Issues Are Divided into Six Categories

Right Advice: Offering a best bundle of products that meet consumer needs. The cost of missing chance: walking in the streets to meet potential customers early without delay.

Time-consuming: Provides quickest claim assistance to loyal customers.

Cost: Increased cost of claims put companies on the losing side in order for them to make a marginal profit [1]. Disruptive models can arise as a result of the IoT, Big Data, and the access to view larger environments than previously possible. These may have a significant effect on the insurance industry across the value chain, altering risk assessment and evaluation as well as consumer engagement models [2].

When insurers invest in digitization, we see advantages for tech giants. We believe the biggest obstacles will be in customer channels and emerging data sources (Internet of Things and Big Data). Investment in core infrastructure would also be needed to allow digitization and cost reduction. Vendors who sell in such areas should see a great deal [2].

The study looks at how technology is starting to reorganize the insurance environment, as well as the industry's future threats and opportunities [3].

6.2 Review of the Literature

The research is categorized into two sections: the power of artificial intelligence (AI), which has emerged as a tech giant, and the obstacles that AI encounters in being a catalyst to the insurance value chain [4].

Difficulties to AI in healthcare emerge/root from a variety of origins taking into consideration the insurance companies, regulatory authorities, and the policyholders. Difficulty in making decisions centers the difficulties of AI. Increasing volume of data is another challenge that has been observed that AI feels difficult to deal with [4].

AI is a computational course that is system-assisted and that aims to create intelligent automated systems. In terms of intellect, AI is categorized as fragile, robust, and bright AI [1].

The report explains the historical problems that have driven various forms of actuarial methods in the literature review sections before going on to review the machine learning (ML) and AI strategies that have been used by analysts in doing such analysis [5].

6.2.1 Artificial Intelligence

The intellect levels of AI are categorized into fragile, robust, and brilliant AI:

Fragile AI, also known as Artificial Narrow Intelligence (ANI), is a form of AI that focuses on a single task.
Robust AI, also known as Artificial General Intelligence (AGI), is a form of AI that mimics human intelligence.
Brilliant AI (ASI5) is an AI capable of innovative and scientific reasoning that surpasses human intelligence [1].

How AI is very smart in global is shown in Figure 6.1.

AI interacts with the atmosphere in the way mentioned in Figure 6.2. AI gathers input sent by the atmosphere and makes analysis based on the input gathered and preceding knowledge. It accepts text in a visual and picture format and analyzes the date to offer powerful solutions [1].

6.2.2 Implementation of AI in Business

For some of our interviewees, we addressed the effects of AI deployment in the workplace. Rearden (2019) said in a personal interview that AI will be

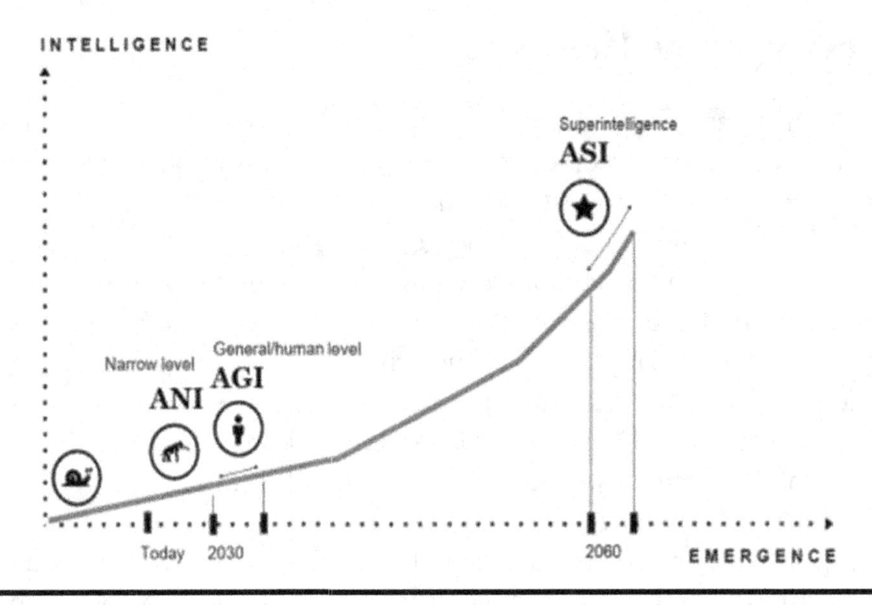

Figure 6.1 How smart is artificial intelligence?

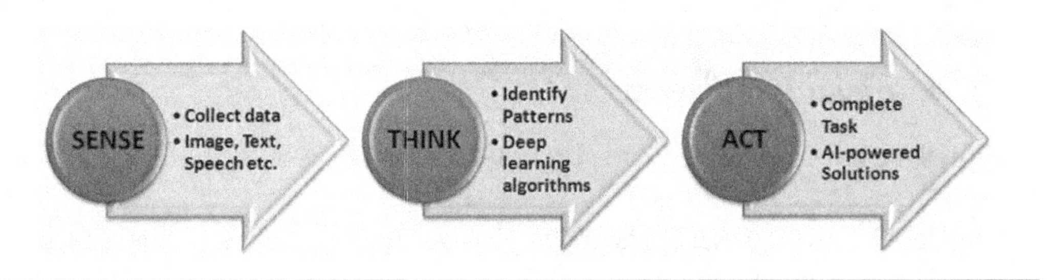

Figure 6.2 Sense-think-act process followed by an intelligent agent.

employed in the future to search the universe of knowledge and suggest the most optimal and socially conscious techniques that create value and tactics to assist businesses in identifying and leveraging strategic plans.

The adoption of AI, according to Rearden, could result in the following outcomes:

- Value generated in society improves brand recognition and attractiveness.
- The value that has been created for workers decreases morale, increases efficiency, and strengthens the company. All of these factors work together and enhance the consumer experience.
- Long-term longevity and higher income.
- Increased chance of successful plan execution and lower likelihood of failure [5].

6.2.2.1 Insurance Business Models

Insurance business models (BMs) can be explained in a variety of ways, including by plotting value chains. Since we live in a data-driven environment, another way to understand a BM is to look at data, its origins, and how it is used. Another factor that affects the prevalence of BMs are platforms such as Google and Alibaba.

There are standard insurers' business managers, who perform a variety of operations on websites such as the billing methods and storage of information [4].

6.2.3 AI in Customer Experience

6.2.3.1 Natural Language Processing

Data mining, subject modeling, and emotion analysis of unstructured web-based social engagement data are all used in natural language processing.

6.2.3.2 Voice Analytics

Call center audio recordings are used to explain when people call and how they feel.

6.2.3.3 Machine Learning

Analysis of the decision tree, learning developed by Bayesian and social sciences are examples of ML techniques that can be used to determine behavior from information [6].

6.2.4 Artificial Intelligence in Claims

Soft robotics: Process mining tools are used to find constraints and increase efficiencies and compliance with traditional claims procedures.

Chart Analysis: Graph or social network analysis is used to detect fraud trends that are present in claims.

Machine Learning: Deep learning methods are used to automatically categorize the sort of damage to cars involved in collisions in order to assess repair costs. To construct claims predictive models, use decision trees, support vehicle machines, and networks of Bayesian.

Internet of Things: Data of IoT is used to create operational knowledge on the occurrence and nature of incidents to offset danger and eliminate damages [6].

6.2.5 Medical Imaging and Diagnostics

The Food and Drug Administration certified an AI Program in April 2018 that tests patients for diabetic retinopathy without requiring expert advice from a specialist IDx-DR, which is a software that accurately classifies patients with retinopathy diabetics 84.7% of the time, shown in Figure 6.3 [7].

6.2.6 Insurance for Self-Driving Cars

Certain collisions, such as those caused by environmental and human causes, will be impossible to prevent even with self-driving vehicles. In six years of texting, self-riding automobiles [8] invented by Google crashed almost and nearly for 11 times as of now.

Self-driving vehicles can pose new risks in a variety of ways, from malfunction of the self-driving process and hacking [9].

6.2.7 Person to Person Insurance

- A person-to-person insurance is one of the insurance models (shown in Figure 6.4), which can be made possible by social media, which allows insurance providers to form online networks to exchange risks.

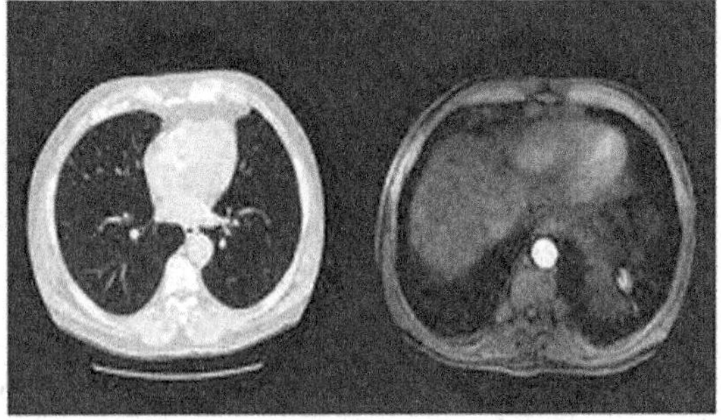

Diabetic Retinopathy

Liver and Lung AI Lesion

Figure 6.3 Diabetic retinopathy using AI.

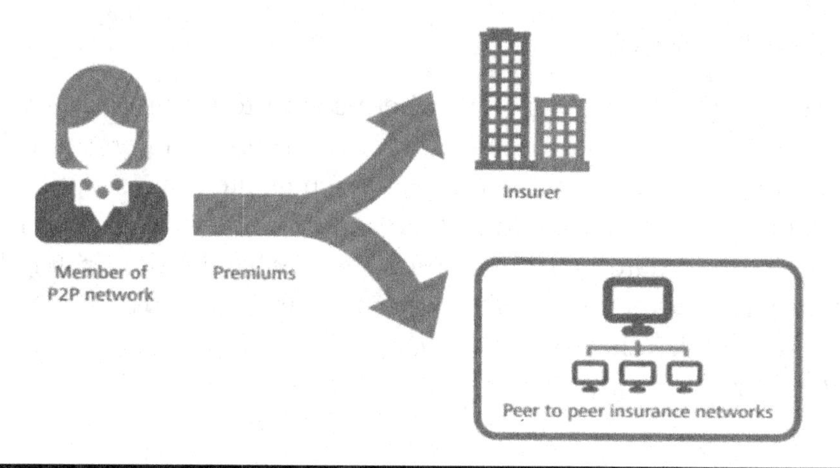

Figure 6.4 Insurance model.

- To share the risk, customers can enter or create their own online social networks.
- Members of the network contribute a part of their premium to a fund for shared benefit.
- The balance of the premium is paid to the insurer by the members [9].

6.2.8 *Insurance for Smart Homes*

Premium discounts are available to State Farm policyholders who install ADT pulse, a home protection device that helps to minimize leakage from a damaged tube, therefore avoiding huge expenditure by getting real-time signals from sensor devices that identify problems of fixtures, among other issues [3].

6.2.9 *Health Insurance*

Google is collaborating with Novartis, a Swiss-based global healthcare provider, for developing a canny wear lenses that try to correct visibility as well as control glucose volumes, which could be life-saving for diabetics [3].

6.2.10 *Insurance for Aerial Assessors*

Drones, otherwise known as lightweight air-borne vehicles, are arousing the curiosity of insurer firms because they offer efficient methods for insurers to enhance business functionalities such as compensation processing and endowing.

6.2.11 Car Insurance

Car insurance is yet another category where connected device can have an effect, as telematics can relay useful data for determining a person's risk level.

In place of depending solely on details such as life time, sex, the type of hooptie someone drives, and old history of accidents, insurance companies can get access to updated information on their policy-holders' riding behavior, such as whether they make sharp turns or sudden stops, determining the speed with which they travel in a car, how often, at what period of a day, and region.

6.3 Research Methodology

Since the future of AI is far beyond what has been observed for now and can build a brand new technological world of healthcare, in locations that reject the prevailing hypothesis, an explanatory approach was chosen. The study examined case lets of healthcare companies and insurers that used AI and performed discussions with veterans [4].

Educational journals, practitioner-focused sources, websites of companies, website of authorities who support the supply chain of insurance companies, parent companies that are owned by insurers, and employees working in an insurance company were among the data sources [4].

Theory was developed using case study research. This inductive method was applied to a total of 20 cases. It was not a random selection; it was focused on how potentially beneficial they are. The cases were selected because they covered a wide range of topics such as incumbents and disruptors, and various geographic areas with various degrees of technological emphasis [4].

Case lets were assessed using the concocts mentioned in the review of literature. The concocts are optimized and validated in a repeated cycle process. Case-by-case review was carried out using the frameworks. One of the writers with an experience in healthcare financials and insurance organizations acted as a "local sinner's lawyer" to boost the study's reliability [4].

Finally, group discussions inviting 12 veterans from major InsurTech companies were conducted. Many of the experts had prior experience with AI implementation in the insurance industry [4].

Their responsibilities include senior executives in insurers' IT departments and insurance review portals, as well as insurance technologies providers and consultants in the field of technology implementation. Overall, they had past experience of 5 years in this industry [4].

The analysis of the current literature on the need for ML and AI in actuarial practice was our primary research approach. To find applicable literature for our study, we created the following structure [5].

Journals not described previously, such as the *Actuarial Journal of North American* and the *Actuarial Journal of South African*, were also mentioned. However, there was no related material in such actuarial papers [5].

This approach has not always resulted in an adequate volume of literature. As a result, we augmented our primary study with:

a. Interviews with people working in the area of AI.
b. Extending the reach of our study to include data from Kaggle [5].

As a basis for picking the sample, this study uses non-probability sampling techniques. Surveys emailed 150 copies of the questionnaire:

1. We have sent mails and fax to few selected and known clients, director and manager of the project, the insurance companies, policy brokers, and advisors of the claim.
2. To Chinese building management scholars [10].

The research paper mainly targets clients, contractors, and insurers of the Chinese economy. The total responses we received sum up to 41. They cover a wide range of regional areas of activity, as well as a wide range of industry experience, company size, foreign market experience, credentials, and contractor styles [10].

This study's research approach is divided into two sections. The evaluation criteria are mentioned in the first step based on the SERVQUAL model's five dimensions. According to the Central Insurance of Iran's 2012 yearbook, 13 insurance companies were chosen for evaluation [11].

6.3.1 We Took SERVQUAL Model to Conduct a Methodology Which Includes Five Dimensions as Follows

1. **Materiality:** Physical factors, new facilities and equipment, personal presence, and the organization's physical environment.
2. **Authenticity:** The ability to provide services quickly and consistently.

3. **Assurance:** Workers knowledge and ability to establish mutual confidence between consumers and service providers.
4. **Responsiveness:** Willingness to provide timely services and to assist customers.
5. **Empathy:** When providing services, service providers pay close attention to the needs of their customers (availableness and comprehending of clients).

The SERVQUAL allows for the assessment of service quality from the viewpoint of the consumer, as well as the tracking of customer preferences and perceptions over time, as well as the differences among themselves.

6.4 Discussion of the Findings

In findings, we'll look at how effective the XGBoost algorithm is at predicting risky clients and potential claims. The efficiency of physical and digital ML classification algorithms to recognize and identify various forms of fraud is then assessed. We are presenting few of the outcomes on implementation of the framework of the block chain [12] in Tables 6.1–6.3.

We used various regression ML algorithms to solve the potential claim since it is a regression problem.
The type of data is considered to be Gaussian in the Naive Bayes algorithm. In addition, for the nearest neighbor, we took the count of neighbors to the same count of classes identified as fraud. We applied equal weighted

Table 6.1 Client Risk Rate (Performance Table).

Classifier	*Accuracy (%)*	*Precision*	*Recall*	*F1-Score*
Decision tree	74.44	0.6473	0.5953	0.6005
SVM	73.21	0.6696	0.5652	0.4841
Nearest neighbor	73.80	0.6696	0.5256	0.4841
XGBoost	**76.81**	**0.6828**	**0.6295**	**0.6392**

Table 6.2 Fraud Detection (Performance Table).

Classifier	*Accuracy (%)*	*Precision*	*Recall*	*F1-Score*	*Training Time (ms)*
Decision tree	92.99	0.870	0.929	0.892	471
Naive Bayes	52.06	0.373	0.520	0.425	**155**
Nearest neighbor	42.70	0.223	0.427	0.255	1254
XGBoost	**99.25**	**0.9928**	**0.992**	**0.9926**	995

Table 6.3 Confusion Matrix

Predicted Observed	*True*	*False*
Positive	True positive (TP)	False positive (FP)
Negative	False negative (FN)	True negative(TN)

uniform distribution. We took the same data for all the classifiers and segregated the train and test data in proportion of 70:30.

To solve the result of the classification, we considered the confusion matrix to calculate the following scores [12].

The attitude of contractors, according to 34.29% of respondents, is an important metric driving the growth of construction in terms of management of risk appetite. In China, culture is found to have a significant impact behind the growth of healthcare management. Chinese people used to be known for their conservatism.

Managing risk capabilities is essential for operations and project management according to 86% of those polled. The people having practice continue catching hold of danger by its cost and nothing to do with attitude of the old system [10] as shown in Figure 6.5.

6.4.1 Recognition of the Face

Face detection is becoming more popular, from unlocking mobiles to boarding planes. As far as facial recognition is concerned, China's unabashed drive for surveillance, along with its AI aspirations, has dominated the news as shown in Figure 6.6.

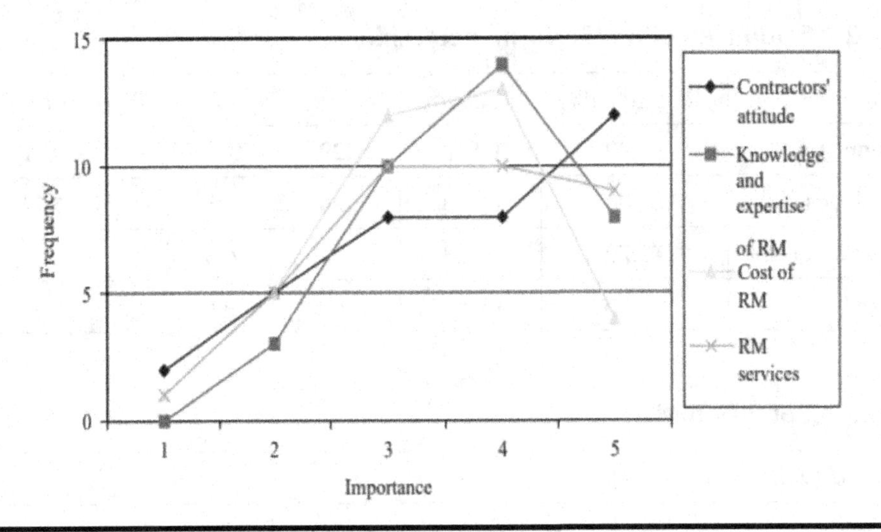

Figure 6.5 Importance versus frequency.

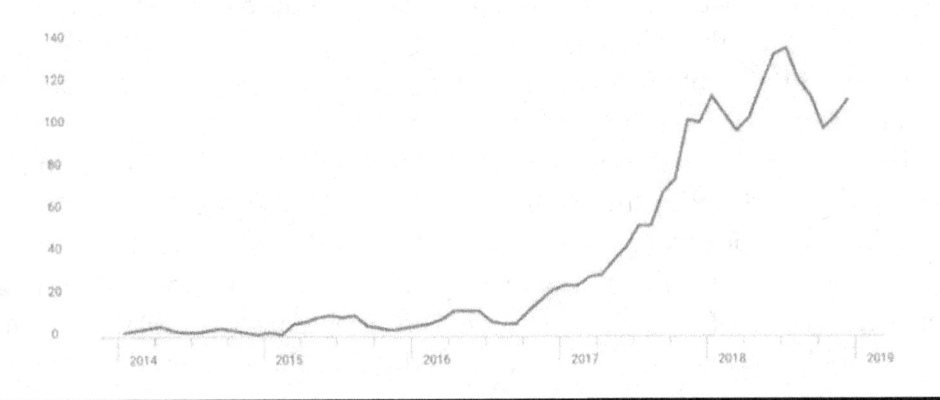

Figure 6.6 China's facial recognition trends up in news mentions.

6.5 Conclusion

Integrating AI in insurance results in improved loyalty of the customer, firm earnings and reduction of the fraud, good time management, and organizational uncertainties. AI use concepts backed by real-life incidents of corporates, demonstrating the vast potential of the insurance sector. New approaches to monitor, manage, and price risk, interaction with consumers, cost cutting, increase performance, expansion of the insurance industry all being enabled by emerging technologies and developments. Back-testing parametric mortality models can be done using ML techniques, especially the RT enhancing system. These methods allow us to identify the flaws in such models using real-world evidence.

Radiation therapy enhancing can be used to predict cause-of-death mortality rates from actual evidence, according to research on cause-of-death mortality in a Poisson model setting. This method makes it possible to spot trends in these odds over time.

This study uncovered information at the business and corporation levels. The focus groups spoke about transformative ability of the AI, which is supported by other findings. Using ML analytics in insurance to develop marketing campaigns, grow the company, increase revenue, and cut costs. The consistency of which statements are predicted will have a direct bearing on the actual economy.

References

[1] Kumar, N., Srivastava, J. D., & Bisht, H. (2019). Artificial intelligence in insurance sector. *Journal of the Gujarat Research Society*, 21(7), 79–91.

[2] Hocking, J., Wood, A., Dally, N., Pan, K., Lin, B., Ban, H., . . . Lee, S. (2014). *Insurance and technology: Evolution and revolution in a digital world*. Blue Paper, Morgan Stanley Research.

[3] Institute of International Finance. (2015). *Innovation in insurance: How technology is changing the industry*. European Banking Authority.

[4] Zarifis, A., Holland, C. P., & Milne, A. (2019). Evaluating the impact of AI on insurance: The four emerging AI-and data-driven business models. *Emerald Open Research*, 1(15), 15.

[5] Literature review: Artificial intelligence and its use in actuarial work. (2021). Retrieved April 11, 2021, from www.soa.org/globalassets/assets/files/resources/res earch-report/2019/ai-actuarial-work.pdf

[6] Yoder, J. (2016). AI in insurance: Hype or reality? [online] *Digital Insurer*. Retrieved April 11, 2021, from www.the-digital-insurer.com/wp-content/uploads/2016/06/716-pwc-top-issues-artificial-intelligence.pdf

[7] Pfeifer, R., & Iida, F. (2004). Embodied artificial intelligence: Trends and challenges. In *Embodied artificial intelligence* (pp. 1–26). Berlin, Heidelberg: Springer.

[8] Hanafy, M., & Ming, R. (2021). Machine learning approaches for auto insurance big data. *Risks*, 9(2), 42.

[9] Hurley, R., Evans, P., & Menon, A. (2015). *Insurance disrupted: General insurance in a connected world.* London: The Creative Studio, Deloitte.

[10] Liu, J., Li, B., Lin, B., & Nguyen, V. (2007). Key issues and challenges of risk management and insurance in China's construction industry. *Industrial Management & Data Systems.*107 (3),382-396.

[11] Saeedpoor, M., Vafadarnikjoo, A., Mobin, M., & Rastegari, A. (2015). A servqual model approach integrated with fuzzy AHP and fuzzy topsis methodologies to rank life insurance firms. In *Proceedings of the International Annual Conference of the American Society for Engineering Management* (p. 1). American Society for Engineering.

[12] Dhieb, N., Ghazzai, H., Besbes, H., & Massoud, Y. (2020). A secure ai-driven architecture for automated insurance systems: Fraud detection and risk measurement. *IEEE Access*, 8, 58546–58558.

Chapter 7

Artificial Intelligence in Agriculture: A Review

Harshitha Sirineni, Thakur Santosh, and S. Deepajothi

Contents

DOI: 10.4324/9781003206316-7

7.1 Introduction

There will be a beginning for every miracle in this universe, where the beginning must link with something before it happens. It may be beyond our imagination, like say "what if machine thinking like a human." This thought in past we called as artificial intelligence (AI) in present. So, let us see the journey of AI from the 19th century to the 21st century.

7.1.1 What Is Artificial Intelligence

An original aspect of AI is reenacting a higher capacity of human brain by programming it to utilize natural language, masterminding speculative neurons where the goal is to shape ideas and furthermore act like people, an approach to decide and quantify issue intricacy, personal development, and arbitrariness and innovativeness. AI isn't a person versus machine saga; it's actually man with machine synergy [1].

7.1.2 Evolution of AI

AI is a well-known concept; its roots go back to the previous time in the ancient Greek era. It was less than a century ago. From then, AI took off high-tech revolution and went from fiction to a very possible reality. In 1884, Charles Babbage worked on an automated machine that will perform intelligent behavior, but he decided that he would not be able to demonstrate the progress of the machine, which he wanted to do, so he dropped his work in the middle. A few years later in 1950, Alan Turing, known as WWII codebreaker and a British mathematician, was universally accepted as being the first person to come up with an idea that a machine can think. He introduced the Turing test, which is used till today, to gauge the thinking ability of a machine in comparison to human. Although his theory at the time was ridiculed, later a proposal was developed in which the term "Artificial Intelligence" was widely known.

In the summer of 1956, Dartmouth College Professor "John McCarthy" coined the term "Artificial Intelligence" and is widely known as the "father of AI."

In 1959, Marvin Minsky, an American intellectual scientist, lightened up the AI lamp amid the darkness of problems; he co-founded the Massachusetts Institute of Technology's AI laboratory. As a result, work on AI continued by Marvin Minsky—one of the leading thinkers through the 1960s and 1970s in the field. He even specified in his book *Stormed Search for Artificial Intelligence* that the problem of AI modeling within a generation will be solved and suggested Stanley Kubrick, a film-maker in *2001: Space Odyssey* was released in 1968, which gained worldwide recognition and became one of the best AI presentations.

In the early 1980s, personal computers came into use, which grabbed everybody's attention and created more interest in machines that think like a human. It took a couple of decades for people to realize the true importance of AI. Nowadays, people are interacting with AI daily in both physically and virtually. For example, smart cars and IoT devices, and voice recognition such as Apple's Siri, Google Assistant, Samsung's Bixby, Amazon's Alexa, and Microsoft's Cortana. Here the common applications are machine learning (ML) and neural networks. Other applications are deep learning and unsupervised learning, which are the future of AI in the creation of high-level thoughts [2].

7.2 Role of AI in Agriculture

The agriculture sector firmly and openly grabbed latest technologies such as AI to improve their productivity and to change overall outcome. AI solutions support overcoming traditional challenges across all sectors in the same way that AI in agriculture improves the flow of information, lowers transaction costs, and speeds communication [3]. These have proven to be the major drivers of economic growth. The agricultural sector discharges have decreased by 20% due to AI changing the method of our food production. Any unexpected natural conditions can be controlled and managed by adapting AI technology.

In today's world, with AI-enabled methods to boot agricultural production efficiency, AI is keeping an eye on startups in agricultural sector. To reduce the rate of unfavorable outcomes, businesses in agriculture are processing agricultural data with the help of AI. Furthermore, AI has improved the applications in agriculture to identify diseases or climate changes faster and respond intelligently, which is currently being implemented to improve agricultural production efficiency [4].

7.2.1 Agriculture in India

The origin of our culture is in agriculture. India's agriculture is composed of many different crops. The agriculture situation is ongoing inefficiency of harvesting, transport, and storage of government-subsidized crops, and environmental disturbances affected one-fifth of total agriculture. Today, we are in the fourth industry revolution of agriculture. All four industry revolutions are described below:

1. The first upheaval depends on mechanical creation gear driven by water and steam power.
2. The second depends on large-scale manufacturing empowered by the division of work and the utilization of electrical energy.
3. The third depends on the utilization of gadgets and IT to additional computerized creation.
4. The fourth is predicated on the utilization of cyber-physical systems.

Compared to other industries, the agricultural sector has been slow to implement, and to empower the fourth industrial revolution, we must adopt varieties of technologies [4].

7.3 Applications and Techniques of AI Used in Agriculture

7.3.1 Crop and Soil Monitoring

AI is used to increase the value of each acre and deep learning for image analysis. It claims to be a self-evolving system with intelligence that gives farming solutions that are future-ready to the agriculture sector. This functioning is carried out by using different IoT-based technology and sensors to monitor plant and health of the soil. Deep learning helps us to identify objects, flora, fauna, and tags them in the image. With this technology, agriculture product grading is done by automated quality analysis from the image, and this alerts on crop infestation and suggests prevention measures [5].

7.3.2 Predictive Agriculture Analytics

AI-based sowing application uses various tools to intimate farmers when to sow seeds, which is the biggest challenge. An advanced feature added to this application is that it indicates risk of pest-based attacks based on climatic conditions and the stage of crop. This information will be sent to farmers in the form of automated text messages or through voice calls [5].

7.3.3 Agri Supply Chain

Agri supply chain [5] is for analyzing real-time data on data streaming that comes from various sources throughout the country. Agri online market place suggests affordable prices for both consumers and producers with the data-driven technology. To ensure the working of Agri supply chain application is fast and efficient, various technology stages were introduced:

- Transition discovery means it analyzes real-time data on various data streams in addition to data obtained by the crowd. This helps us to discover automatic transaction to make high margin for the benefit of producers and consumers at the same time.
- Quality maintenance refers to automatic sorting and grading on AI-based and computer vision for trading good quality across country.
- Credit risk management handles the credit default problem that rises in supply chain to reduce operation risk rate.
- Agri mapping analyzes images that are captured by satellite, gathers information with the help of deep learning, and obtains maps in real time at high resolution.

7.3.4 Drones

Drones are capable of in-depth field analysis, long-distance spraying, and efficient crop monitoring. All these works are done by sensors that capture images and videos; identify nutrient deficiency, pest damage, fertilizer needs, water quality, and chemical composition of surface; and track the surface temperature and lidar sensor shows farm in 3D model. Compared to most of the farmer's equipment, drones are affordable [6].

7.3.5 AI Agriculture Bots

Agriculture bots are AI-enabled and used for helping farmers in order to protect the crop from weeds in the most efficient manner. This can be done by monitoring weeds with the help of computer vision and spraying them. By using this application, we can harvest high volume of crops at a faster rate than human labor [7].

7.3.6 Precision Farming

The goals of precision farming are profitability, efficiency, and manageability. There are key innovations that empower accuracy in cultivation: high exactness situating framework, mechanized guiding framework, geo planning, sensor and far off detecting, incorporated electronic correspondence, and variable rate innovation. These technologies provide advice about crop rotation, water management, nutrient management, optimum planting and harvesting time, pest attacks, and many. Precision farming can be summed as a "Right place, Right Time, Right Product" [6].

7.4 Future of AI in Agriculture

Assuming you need to take care of the world in the future, we'll need to deliver a same amount of food in the next 40 years as we did in the previous 8,000 years, which shows a hint of the food framework's conflict. With the development in populace, change in utilization conduct, and environment emergency, how would we get our food creation?

The genuine mystery is the manageable creation. It ought to be with less data sources, less composts, less pesticide; and less water everything should be manageable. Else we will annihilate our plants. The security of the food framework is one of the world's most squeezing difficulties. It all relies upon how we tackle it; if everybody on earth is on diet of the normal American, then that would require all tenable land to be utilized for horticulture, and we are as yet being 38% short. Also, if that is correct now, then what do we do when there are two billion additional individuals?

All things considered, the key is surprisingly energizing, and that is proficiency. Essentially, how would we deliver significantly more on the land that we are as of now utilizing? At the point when we need to go for feasible creation, everyone should be a part of the modified framework and embrace development to achieve our goal of making proficiency in your work that is unrivaled anywhere else on the world.

That strategy is vertical cultivation; those are distribution centers with piles of tank-farming systems to improve the growth of verdant crop. They are being developed in urban communities everywhere in reality where new produce and land are scant. The critical deterrent here is the expense of energy and the cost utilizing a ton of it takes on the climate. The potential gain is that counterfeit lights and environment-controlled structures permit developing day and night, all year, creating an altogether better return for every square feet than an external farm. For now, however, just costly, verdant greens or spices have shown high yield in the vertical framework. In addition, the jury is certainly still at a loss as to whether this is frequently really harmless to the ecosystem method; potential arrangement is made to use blue and red light frequency of photosynthesis development and a high growth rate. Another development in the home is open farming work, which means to make "inventory of environments" so temperature and moistness can be set to reproduce the ideal conditions for developing harvests that may ordinarily come from wherever in the planet, locally all things considered. This is frequently a push to handle the food miles issue [8].

When the delivered are dispatched all throughout the planet, it makes pointless CO_2 emanations. What's more, "Our Future is about Accepting Challenges."

7.5 Challenges in AI Adoption in Agriculture

The Agri input companies face many challenges when they try to manage a demonstration farm; these farms are known as Demo Farms. Nowadays, Agri input companies utilize these farms in various regions to showcase new seed varieties or agro chemicals or to demonstrate new Agri culture techniques in parallel to the existing traditional ones. Let us see some challenges that Agri input companies face: collection of data across regions is a slow and long process, and the manual work gives more chances to errors such as loss of information. Because of that only pen and paper spread sheets or elementary online tools are being used for data acquisition and processing. So, what can be done to boost farmer loyalty with dozens of competitive products battling for farmer attention. Agri business companies are also facing difficulties in retaining the loyalty of farmers. Designing effective farmer engagement programs is also becoming very difficult in the absence of reliable farm-level intelligence. The lack of real-time data to understand the factors that impact crop growth such as climate change and resulting weather conditions, pests, and diseases, availability of soil nutrients, and water stress level affects agricultural productivity. Is there any comprehensive solution for these problems?

Yes, Agri input companies give accurate ground data, schedule daily task, manage farm activities, and much more like looking for weather- and satellite-based advisory to stay on top changing conditions and help farmers to adopt a smart package of practices that reduces production costs and improves efficiency. For the management team, they provide access to customize farm reports' ability to provide real-time guidance on field requests and measure staff performances. The entire crop production cycle monitors health and accurate yield estimations to make sure that farmers are ready for sustainable and climate-resilient farming [7].

With all these challenges if we add up the most powerful challenge, that makes scale productivity, increases efficiency, and strengthens sustainability. That challenge is image recognition to identify pathogens pest infestation in crops. If we can identify them early enough, we can spray fewer chemicals that lead to less crop distraction, leading to bigger crop yields that could mean more food. To make this challenge happening, we need to build a better database. These are the challenges AI faces in the field of agriculture. "Challenging for Good, we see a Better Future."

7.6 Limitations

It costs a lot of money to make or buy the technology. Indian farmers are already in debt, and because of this few of them are taking their own lives; hence, bearing the cost of the new technology will not be an easy task for the farmers.

Technologies need maintenance to keep them running. Thus, such maintenance is another additional cost for the farmers. Furthermore, an Indian farmer holds less than 5 acres of land, so the output will be limited and taking out all the expenses will be difficult.

By advance technology in farming, farmers can lose their jobs, and in India, around 70% of the population depends on agriculture directly or indirectly, so this will push most of the people into poverty.

Poor farmers are denied access. As a result, roughly 70% of farmers have land that is smaller than 5 acres, making it incredibly difficult for them to implement the technology.

7.7 Conclusion

AI complies with an important role in agriculture that moves toward more automation and acting on real-time applications with an accurate system in place. Precision agriculture came into picture, instead of traditional agriculture, with its low cost and advanced tools and equipment. For this, applications need to be more robust so that the scope in agriculture lives on and miracles are made out of it.

Farmers are displaying signs of confidence of a potential future by adopting cognitive solutions that handle realistic challenges. This leads to people becoming capable of buying sufficient food. All these applications and techniques develop technology to make everybody live better on this earth. The agriculture land is farmer's legacy, and "farmer is the only person in our economy who buys everything at retail, sells everything in wholesale, and pays the freight both ways," this was the statement by John F Kennedy [9]. If we don't talk about a farmer, there will be no perfect meaning for agriculture. Farmer is a "Character," if *character adds on with intelligence—that will be the goal of true and future Agriculture.*

References

[1] Marr, D. 1977. Artificial Intelligence: A Personal View. *Artificial Intelligence*, 9(1): 37–48. doi:10.1016/0004-3702(77)90013-3
[2] www.businessinsider.com/artificial-intelligence?IR=T
[3] https://builtin.com/artificial-intelligence
[4] Dharmaraj, V., and Vijayanand, C. 2018. Artificial Intelligence (AI) in Agriculture. *International Journal of Current Microbiology and Applied Sciences*, 7(12): 2122–2128. doi:10.20546/ijcmas.2018.712.241
[5] Sharma, R., Kamble, S. S., Gunasekaran, A., et al. 2020. A Systematic Literature Review on Machine Learning Applications for Sustainable Agriculture Supply Chain Performance. *Computers and Operations Research*, 119: 104926.
[6] Shafi, U., Mumtaz, R., García-Nieto, J., Hassan, S. A., Zaidi, S. A. R., and Iqbal, N. 2019. Precision Agriculture Techniques and Practices: From Considerations to Applications. *Sensors*, 19(17): 3796. doi:10.3390/s19173796
[7] Pandey, A., and Srivastava, M. 2019. A Study of AI Agents in Agriculture—Present Application & Impact. *Amity Journal of Computational Sciences (AJCS)*, 3(2), ISSN: 2456–6616 (Online).
[8] Nawaz, A. S. N., Nadaf, H. A., Kareem, A. M., and Nagaraja, H. 2020. Application of Artificial Intelligence in Agriculture-Pros and Cons. *Vigyan Varta*, 1(8): 22–25.
[9] www.financialexpress.com/opinion/if-farmers-remain-poor-so-will-the-country-here-is-why-and-how/1035532/

Chapter 8

Machine Learning and Artificial Intelligence-Based Tools in Digital Marketing: An Integrated Approach

Preetha Mary George, Sanjeev Ganguly, and Venkat Reddy Yasa

Contents

DOI: 10.4324/9781003206316-8

8.1 Introduction

Artificial intelligence (AI) and machine learning (ML)-based algorithms have transformed the business and in different working strategies of the organizations in each and every sector. The technology has improved rapidly in recent years, which has created a completely dynamic digital marketing environment. About 80% of available data on the internet (Big Data) has been generated in just few years [1]. The interaction and communication among the companies and its customers has now become very easy due to the use of such AI-based technology. Also when it comes to the marketing perspective, AI has completely made it unique and more individualized (digital marketing based on AI is more focused to address the client individually).

Data present on social media sites, search engines, and advertising and shopping sites is huge, and thus, companies are not investing more and more to improve the ML ability and to improve the digital marketing strategies. Many organizations from various sectors nowadays use online marketing so as to boost business and to build a good marketing strategy [2].

We know that it is very difficult to interpret such huge data and to draw insights. It will take more time and could possibly involve many errors, which can lead to draw wrong conclusions. So many analytical tools have been used by the companies these days, so as to increase the speed of work, to remove the chances of error, to optimize and systemize the processes, and to reduce human intervention—automate the work.

The technologically advanced ML tools are now being used by most of the sectors so as to learn and use the past activities and historical data of the customers and thus help in the digital marketing strategies of the organizations.

One of the major reasons why companies are trying to shift to AI- and ML-based analytical models is because of the complexity in using the basic statistical tools. The basic analytical tools used by the managers were complex; very little was known about these tools and also human intervention was more as compared to AI-based analytical tools. But when it comes to the overall usage of this technology in digital marketing, it is very less, that is, the use of AI and ML is still at its starting stages only. This research mainly explores that—up to what extent the managers or top management are aware about these tools. The perceptions of the company's top management concern were about AI and ML and these tools; the degree to which they are currently using these tools; and how to adopt and use these AI-based tools in their strategic decisions, especially in digital marketing [3].

8.2 Literature Review

8.2.1 Key Concepts

Artificial Intelligence—Human intelligence is simulated to machines, which makes the machine to respond or act intelligently as a human. In simple words, we can say that AI is basically making a machine to behave, respond, or interact (in any situation) in the same way as humans do. An artificially intelligent machine possesses all the similar traits to human mind so as to respond in a similar fashion (as humans do). A perfect example is the face recognition function in mobile phones.

Machine Learning—ML is particularly a part/subset of AI, where system learns and automatically with no human intervention involved. ML basically involves developing computer programs, which can use the data easily to learn for themselves. Some of the techniques of ML are supervised ML, unsupervised ML, semi-supervised ML, and reinforced ML [1].

Big Data—It is a field which deals with the analyzing, revealing various information, trends, and interpretation of extremely large amount of data sets that are practically impossible to be dealt by the traditional methods of analyzing data. Examples include users comments, reactions, reviews on social media about any particular company, brand, or a product [4].

Previously, the strategic decisions made by the organizations were dependent on the data gathered internally to the organizations which were the past data only. But nowadays, the decisions are taken while predicting the future, so as to gain a competitive advantage in the market. By the availability of new AI-based analytical tools and huge data (Big Data), it has become more easy for the companies to analyze and know about the consumer purchase behavior, reviews about any particular product and thus to make more effective, long-lasting strategies, and faster decisions. Also, it is easier for the companies to know about the success rate of the current strategies implemented by them in the field of digital marketing. Selection of the analytical tool is totally dependent on the type and the complexity of the business.

These AI-based analytical tools in digital marketing can:

i. easily access Big Data,
ii. clearly structure the data,
iii. help in making effective and faster decisions,
iv. target individual customer's needs,
v. clearly indicate the company success factors,
vi. help in engaging with the customers more easily,
vii. help the managers to target new customers easily, and
viii. adapt the digital advancement very easily.

New integrated skills actually boost system performance constantly. ML consists of procedures through which a computer learns (similarly as humans). One of the best examples of AI is a chatbot on a website. This not only will boost the ability of client support department but also increases automation with no human intervention. Various studies have also claimed and showed that data mining, ML, and text mining are used to analyze the behavior of customer on a website. Some literatures showed that many new methods have been developed, to extract the information about the user behavior and intentions—just by analyzing their posts, comments, and their reactions. These techniques help the marketers and managers to get a deeper insight about users and to develop a better understanding about their customers, which helps them to make their advertisement content accordingly for different segments [3]. Getting insights about the users' intention from online platforms or social media sites has very good response for digital marketing field. Conditional random field—which is an advanced graphical model, used for sequence data has now been developed, and thus, these things have taken the ML models to greater heights, in digital marketing.

While doing the literature review, we came to know about various studies which showed some of the biggest advantages of analytical tools (AI based) in digital marketing. Also with the introduction of new tools of analysis, there are both challenges and opportunities in the area of online/digital marketing.

Finally we can say that:

i. By the use of these tools, machine performs optimally and steadily.
ii. Calculation is much faster while performing analysis, and thus, these tools enables us to make better and faster decisions.
iii. Complex analysis can be done very easily with vey less chance of error.

iv. Managers also get insights about those users, areas, and segments (for digital marketing), which are unexplored by the organization.
v. Routine activities can be optimized easily, and there is less or no chances of error as human intervention is not there while performing with these AI- and ML-based analytical tools—research aim.

The aim of this research study is to identify the reasons—why these AI-based tools are not used frequently at a good speed by the organizations. Secondary aim of the study is:

i. To know about the perceptions of the management about these AI- and ML-based analytical tools.
ii. To identify the level of awareness among managers, with various terms related to AI and ML.
iii. To know about the AI- and ML-based tools used by the managers and how to adopt ML in digital marketing.
iv. To know about the role these tools play in decision-making and in setting up digital marketing strategies.

8.3 Methodology

Ongoing through various research papers, we found that most of them have used questionnaires and conducted in depth interviews in order to collect the responses from the respondents. Convenient sampling was used for selecting the sample size, and most of the respondents were the higher ranked employees (departmental heads) or top management employees/managers in the organizations (organizations which are involved in digital marketing). Employees at executive level were also selected by the researchers in some of the studies. For in-depth interviews, experts from marketing and strategy development department were chosen as they have full knowledge about which analytical tools company is using for analyzing the online data.

Questions in most of the studies generally include: (i) general questions—just to collect basic information about the respondent and to know whether he/she is eligible to answer the questionnaire. (ii) Some questions were included to check the usage of analytical tools in the organizations, so as to know about managers' perceptions about these AI- and ML-based analytical tools. (iii) Awareness level questions—just to check the level of awareness with various terms associated to AI and ML. (iv) Practical applications questions—they were included just to collect the respondent's opinion about—how these tools help in decision-making, practical applications of ML in digital marketing and in setting up digital marketing strategies. Also in the questionnaire, the questions included were a mix of rating scale questions, closed-ended questions, and open-ended questions. Most of the questions were closed-ended, while a few questions were based on the Likert scale. But when it comes to open-ended questions, we observed that most of the researchers have only included one open-ended question or maximum two open-ended questions. As analyzing these open-ended responses is a difficult task, they just kept these questions to get only the respondent's opinion about any particular area or topic.

So, these types of questions were asked by the researchers, and then conclusions were drawn based on the responses.

8.4 Findings

Based on the information obtained, the findings of the research paper are as follows.

8.4.1 The Perceptions of the Management about the Analysis Tools

The main intention is to find out how respondents distinguish market analysis and analytical tools. The intention was also to know how to use the tools for their online marketing processes. The questions that were asked to the respondents were broadly on the basis of the market analysis that they do for planning process and implementing online strategies for marketing. All the information (activities) on the internet is measurable with the help of digital technologies, and this information is collected in making effective decisions for their growth. Managers are very much aware about it, and they also use analytical tools in order to measure and analyze online content. From the responses, it is clear that analysis is important as it aids in providing sources of information. These information sources are then again used to prepare and implement the digital marketing strategies. There are many analysis tools present in the markets that can be used for improving DM strategies and to gain advantage over the competitors. Some of these online analytical tools are Google Analytics, Google Data Studio, Facebook insights,

Facebook ad manager, etc. [5]. The aforementioned tools can be integrated for the preparation of online marketing strategies. Research papers suggest that the list of analytical tools is used mainly by agencies, media houses, and advertisers. The usage of analytical tools is completely based on the business needs. From the recent research papers, we get to know that a large section of people believe that it is important to use data analytical tools in digital marketing, and Big Data plays a vital role in online marketing. Marketing managers strongly believe that in today's business, ML tools play a major role.

8.4.2 Awareness Among Managers/Top Management with Various Terms Related to AI and Ml

From research papers, we get to know that respondents have rough knowledge about AI and Ml and also Big Data. They understand Big Data as a huge amount of information than can further be processed. People also feel that AI and ML along with Big Data create competitive advantage and is the future of digital marketing [4]. A convenient number of people think that data management is important and additional education is required because one needs to have sound knowledge to embrace the opportunities in this field. From the research papers, it is clear that respondents use analytical tools to extract useful information from sales reports and historical marketing campaigns. Few gather data from third-party sources and sometimes free available database. People also feel that AI and Ml and Big Data usage in digital marketing have few obstacles, and they are:

(i) Most people think that planning and implementing platform data management is extremely high.
(ii) Implementation of data-oriented approaches is time-consuming.
(iii) Digital marketing is a relatively new platform, and integrating Big Data requires a lot of information; this puts pressure on educating and training employees.

When it comes to AI and ML, research papers clearly say that people are aware about the terms and they feel that these are applicable in present time. People identified ML as a kind of process that requires huge amount of information for any machine to learn and adapt. Many people also agreed that AI and ML have huge potential in this technology-based DM field.

The process basically involves working with huge/large data (gathered from websites). Respondents feel that the analysis part will mostly be done by software and machines, while strategic decision-making and creative work will be carried out by humans.

8.4.3 Tools Used by Managers (Based on AI and ML) and How to Adopt ML in Digital Marketing

Research papers gave us an idea that people have basic knowledge about the AI- and MI-integrated platforms such as Google Data Studio, YouTube, Google ad words, Facebook ad manager, and Mailchimp. People also know about the automatic text translators, automatic AI-based cars, and graphics software/card. We have gone through handful number of research papers, and we got to know that ML will affect some of the areas in online marketing/DM [6, 7]:

i. ML will affect advertising and also the management of advertising campaigns.
ii. ML will automate the reporting process.
iii. ML will reduce human intervention in the areas where partial automation is required for example—email or chat communication.

Above are the mentioned areas discussed where ML will minimize human effort, but areas of creative process and building and maintaining relationship with partners will limit ML usage. ML-integrated softwares can develop ideas and also draw from pictures, but the result is not close to what humans deliver, particularly because in machines, emotional factor is missing.

People feel that machines are likely to replace the humans in various sectors, but there are still many sectors where it will be difficult for machines to control the situations and manage strategically [8].

8.4.4 Role of AI- and ML-Based Analytical Tools in Taking Decisions and in Setting Up the DM Strategies

Research papers suggest that the framework that aims at encouraging the use and adoption of these ML- and AI-integrated analytical tools in digital marketing comprises two things/factors:

i. Management and organizational culture that will create the atmosphere where the projects will be carried out and successfully finalized.
ii. A map of the project containing four phases and they are:
   a. Management support—without top-level managers' support, AI and Ml tools cannot be utilized properly. Top-level employees should first lead and must tell the importance of analytical tools and its contribution toward business success [9].
   b. Out-of-the-box solution—a company's culture should motivate employees to use analytical tools for innovative solutions.
   c. Technical infrastructure—every company should have a team that has technical knowledge for everyday operational needs. It is an advantage for a company that has an IT team or has good relationship with technical partners and developers.
   d. Focus on data analysis—it is important for the companies to do a detailed analysis of the data available as it will help them to deploy their amenities in a better way to gain competitive advantage through their DM strategies and will also help them to boost internal management [5, 10].

8.5 Conclusion

After reading research papers in depth, we get to know that market analysis is important, as it helps to take better strategic decisions. It is also confirmed that people depend on market analysis tools and analytical tools before making a digital marketing plan. These analytical tools, used by organizations, have boosted their growth trajectory. Upon reading the research paper, we got to know that analytical tools can help companies in certain ways, and they are

a. help in understanding competitor activities and their strategy;
b. aid in important internal data into third-party analytical tools;
c. aid in visualizing of data;
d. aid in quick access to data and information; and
e. help in tracking real-time data for ongoing campaign.

After understanding the research papers, we conclude that the usage of AI and Ml will be of great in the field of digital marketing. It will also help companies to use quality data and process automation. There is no doubt that AI and ML have the place of digital marketing. Despite its hurdles, marketing managers believe that AI and ML are the future of DM. But in the current scenario, it is clear that there is still room for adoption of advanced technology in the workplace despite its availability.

References

[1] Ma Liye, and Baohong Sun, "Machine Learning and AI in Marketing–Connecting Computing Power to Human Insights", *International Journal of Research in Marketing*, Vol. 37, No. 3, pp. 481–504, 2020.
[2] Geng Cui, Man Leung Wong and Hon-Kwong Lui, "Machine Learning for Direct Marketing Response Models—Bayesian Networks with Evolutionary Programming", *Management Science*, Vol. 52, No. 4, 1 April 2006.
[3] Hair Jr, Joseph F., and Marko Sarstedt, "Data, Measurement, and Causal Inferences in Machine Learning: Opportunities and Challenges for Marketing", *Journal of Marketing Theory and Practice*, Vol. 29, No. 1, pp. 65–77, 2021.
[4] Pal Sundsoy, Johannes Bjelland, Asif M. Iqbal, Alex Sandy Pentland and Yves-Alexandre de Montjoye, "Big Data-Driven Marketing: How Machine Learning Outperforms Marketers' Gut Feeling", *Social Computing, Behavioural-Cultural Modelling and Prediction*, MIT Open Access Articles, 2014, pp. 367–374.
[5] Marketing Analytics with AI Complete Guide, https://research.aimultiple.com/marketing-analytics/
[6] Joni Salminen, Yognathan Vighnesh, Juan Corporan, Bernard J. Jansen and Soon-Gyo Jung, "Machine Learning Approach to Auto Tagging Online Content for Content Marketing Efficiency—A Comparative Analysis between Methods and Content Type", University of Bradford, 2019, Elsevier.
[7] Tim Mackey, Janani Kalyanam, Josh Klugman, Ella Kuzmenko and Rasmi Gupta, "Solutions to Detect, Classify and Report Illicit Online Marketing and Sales of Controlled Substances via Twitter—Using Machine Learning and Web Forensics to Combat Digital Opioid Access", *JMIR Publications*, Vol. 20, No. 4, 27 April 2018.
[8] Vinicius Andrade Brei, "Machine Learning in Marketing—Overview, Learning Strategies, Applications and Future Developments", *Foundations and Trends in Marketing*, Vol. 14, No. 3, pp. 173–236, 31 August 2020.
[9] Louis Columbus, "10 Ways AI and Machine Learning Are Improving Marketing in 2021", *Forbes*, 21 February 2021.
[10] Myron Monets, "How Artificial Intelligence and Machine Learning Can Be Used in Marketing", CURRATI—Editor of Chaos, 23 April 2018, https://curatti.com/ai-machine-learning-marketing/

Chapter 9

Application of Artificial Intelligence in Market Knowledge and B2B Marketing Co-creation

H. Raghupathi, Debdutta Choudhury, and Cynthia Jabbour Sfeir

Contents

DOI: 10.4324/9781003206316-9

9.1 Introduction

Businesses have often benefited from technological advances because they have provided new ways to reach out to consumers. Artificial intelligence (AI) is one of the most important innovations of our time, and it is causing quite a stir in the modern world [1, 2]. AI in B2B marketing is here to change the way people communicate with brands, content, and services because of its marketing ability.

AI in B2B marketing will benefit not only companies, but it will also benefit consumers by inspiring them and providing them with more than they can imagine [3].

Effect of AI on business:

- Centered on cognitive technology outcomes and makes faster business decisions.
- AI programs can minimize errors and "human error" if they are well set up.
- Use data to anticipate consumer needs and have a more personalized experience.

AI marketing based on collection of data, study, as well as additional audience or economic dynamics findings can affect marketing activities. AI is often used in marketing campaigns where speed is significant. AI can also provide industrial firms with various forms of market intelligence that are essential for B2B marketing [4]. Learning a person's surroundings and behaving in order to accomplish goals are only a couple of the mechanisms involved in successful intelligence adaptation [1].

In this paper, we briefly overview that the theoretical changes of marketing implemented have resulted from technological advancement in the digital environment, especially in the B2B context [5].

9.2 Literature Review

Technology is evolving at a rapid pace, and organizations are keen on leveraging this technological advancement in order to help generate more revenue and propel the business forward. B2B organizations however have relied heavily on the conventional mode of communication, the communication takes place through multiple organizing interactions between the buyer and seller. These interactions are meant to build trust factor for the longevity. The agenda is to influence the buying behavior of customer with the help of comprehensive communication strategies [5].

Organizations are entering digital media space to help improve the communication with its prospective clients and social media analytics tools, and AI is ubiquitous in this predicament [6] AI helps in negating crisis and improves the ability to co-create. The machine learning (ML) algorithms play a significant role as they help in understanding the sentiment of the supplier or buyer while conversing through a B2B market place such as IndiaMART. The ML algorithm uses sentiment analysis to understand the behavior patterns of the parties involved in the process [7].

The e-commerce platforms have reshaped the B2B organization space and redefined the way they communicate with the buyer and supplier. The interactions are primarily confined to the chat on the platforms and then they are taken into telecommunication on as and when required basis. This provided a huge scope for AI to reduce the work load on people by analyzing the chat process to detect patterns and create FAQ's for the company [8].

With the world being as it is under this new normal, the B2B organization interactions in the physical space has been reduced tremendously which opens doors for digital B2B interactions and this predominantly happens over platforms like zoom and Skype, and organizations are creating AI tools so that buyers can interact with it and customize the service offerings to their needs [9].

9.3 Artificial Intelligence and B2B Market

Many areas of B2B marketing are being revolutionized by AI. Gartner (2018) predicts that by 2020, to drive marketing automation, one-third of all B2B marketers would use AI-enabled tools. Thus, using data relating to past user information and buying patterns within the B2B business's website, AI helps to improve customer connections. This is basically a next-generation customer experience model. It also opens plenty of expansion and cross-selling possibilities.

Chatbots are another way AI is used in marketing. These bots can assist with problem-solving, product or service suggestions, and sales support. Marketers benefit from AI because it analyzes data on customer behavior faster and more accurately than humans.

Most of us have seen the advantages of AI in our very own experience of shopping, as in when we use Uber to get a ride, or when we use Netflix to find a movie, or when we use Airbnb to book a room. Many conventional balance sheet and operating income line items are much less predictive of financial viability than effective use of Big Data and data science.

Over the past ten years, marketing departments have used social networking, customized website creation, content marketing, account-based marketing, marketing analytics, targeted advertisement, marketing and sales coordination, programmatic advertisements, attribution models, and funnel metrics [10].

Marketers must collect, analyze, and identify data before targeting possible leads using conventional data generation methods, which take a significant amount of time. While marketers can generate leads using data from a variety of websites, CRM systems, and marketing campaigns, there are some gaps in which marketers fail to understand the buying patterns and interests of their leads.

AI can easily capture a broader range of data from different sources as blogs, social media platforms, and contact databases due to its wide coverage. AI can close the gap between salespeople and potential customers by determining the best B2B goals for inbound and outbound marketing campaigns [11, 12].

9.4 Challenges of Artificial Intelligence

B2B consumers, for example, are often annoyed by irrelevant and intrusive ads distributed by automated email and social media marketing with AI. AI-created unsuccessful marketing campaigns have the potential to damage a company's brand and credibility.

A solid IT infrastructure is needed for an effective AI-driven marketing strategy. AI technology processes massive amounts of data, which can be costly to set up and maintain. They'll almost certainly need regular updates and repairs to keep running smoothly. This can be a major stumbling block, particularly for smaller businesses with limited IT budgets.

Although large corporations can choose to create and operate their own AI marketing tools, smaller businesses may benefit from cloud-based solutions.

Even though AI solutions usually have a high return on investment, a business argument must be made to justify investing in these new technologies. This is especially challenging in smaller businesses with already tight budgets. Complex software and high-performance hardware are needed for AI technology, which is costly to implement and maintain.

Organizations no longer have to rely on designing in-house AI solutions, due to an increasing number of affordable AI vendors. AI marketing technologies can be applied not only more affordably, but much far more quickly than in the past.

There is still an AI skills shortage, which can have a significant effect on companies looking to implement AI marketing strategies in-house. To fill these new vacancies, the existing pool of AI talent is not rising quickly enough. Although some companies may be able to close the skills gap by educating current staff, others may need to set aside funds to recruit AI specialists with a competitive compensation package.

For regulatory purposes, certain companies may be prohibited from storing data offsite, which means they may be unable to use cloud-based AI marketing vendors. All companies must take responsibility for ensuring that AI software is used safely and in a way that benefits their consumers rather than just their bottom line.

9.5 Methodology

Mixed methods research was followed to analyze the data, where qualitative research was conducted on North American-based startup company which provides competitive intelligence services to the customers using AI technology backing with ML and natural language understanding (NLU). Senior employees from various organizational levels are participated on semi-structured interviews with a time slot of half an hour [1]. Further quantitative research used panel data and common system method of moments caused by an independent variable that is related to the error word [5].

Population: Managerial level employees from various departments including sales, customer service, marketing, and IT-related people participated on the in-depth semi-structured interviews [1].

Sample Size: A startup company employees based on North America are considered as the sample size where they service competitive intelligence services and corporate sales to the clients using AI.

Data Sources: The initial step was data drafting on B2B sales and modifying with the suggestions from the colleagues before conducting the interviews. After conducting couple of interviews and taking feedback on questionnaire further modified based on the suggestions, this process was repeated two times, and after the final best questionnaire, in-depth interviews are conducted to analyze the data on B2B sales and value creation by AI in B2B marketing [13]. The complete interviews time was 7.25 hours with an average time of 30 minutes for each, whereas the range is between 16 and 40 minutes on overall [1].

Interviewer wrote a summarized memo and submitted the audio recordings transcribed using professional service of Microsoft word. These summarized memos and audio transcripts are the data set of this study [1].

9.6 Findings

9.6.1 AI in Co-creation of B2B

Initially, when start-up firms want to interact with its potential clients, they develop an AI application (basically a bot) [6, 7]. This AI bot follows the below process to incorporate sales of IT products and services is shown in Figure 9.1.

9.6.2 B2B Funnel

There are seven steps in AI as mentioned below [8, 9, 14]. This method has been made easier thanks to AI. The following sections detail how AI influenced B2B revenue is shown in Table 9.1.

9.6.3 Market Knowledge in B2B

Only when various types of AI frameworks and inputs from various types of industry expertise are combined can high-quality innovation be produced [2, 15]. Customer experience, consumer knowledge, and external business knowledge are examples of market knowledge [4, 16].

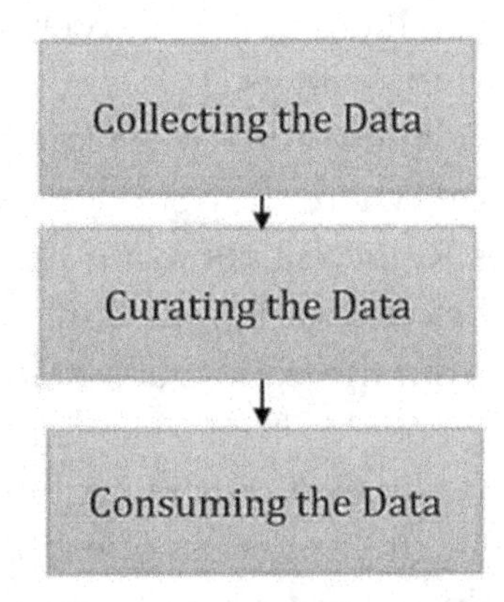

Figure 9.1 AI workflow in co-creation B2B process.

Table 9.1 How AI Contributes to Processes in B2B Sales Funnel

Process	*AI Contribution*
Prospecting	Lead qualification and generation Models
Pre-approach	Advertising, retargeting
Approach	Contact through digital Agents, content curation
Presentation	Prototyping, sentiment analysis
Overcoming objections	Competitive intelligence
Closing	Dynamic pricing according to client
Follow-up	Chatbots, automated workflows

Table 9.2 How AI Contributes to Different Types of Market Knowledge

Type of Marketing Knowledge	*Contribution of AI*
Customer knowledge	• Creating profile of potential customers. • Structured and unstructured data about customer attributes like buying behavior and buying process. • Predictive models for prospect scoring • Chatbots
User knowledge	• Internet of things • Big Data • Text analysis • Social media analysis
External market knowledge	• Natural language processing • Competitive intelligence • Analyzing press releases and blogs for insights

Table 9.2 gives a brief overview about how AI contributes to market knowledge.

9.7 Conclusion

Nowadays, due to technological advancement, many segments have started using AI and ML as their base. AI is showing a huge impact on present marketing activities [1]. The previous results have discussed about how AI is contributing for co-creation and marketing knowledge in B2B marketing and sales. AI also reduced the work of marketing or sales professionals in this vast ocean of B2B marketing [3]. The previous results also show the usage and importance of AI-based sales funnel. In this summary, the differences between traditional and AI-based systems are discussed.

By adopting strict concords, dependable templates, and a personalized approach, value co-creation is encouraged by AI-based providers [4]. The use of six AI building blocks in B2B marketing was also addressed previously [17]. Overall, the summary discussed the uses and usefulness of AI in the B2B field for co-creation and business awareness.

References

[1] Paschen, J., Kietzmann, J., & Kietzmann, T. C. (2019). Artificial intelligence (AI) and its implications for market knowledge in B2B marketing. *Journal of Business & Industrial Marketing*, 34(7), 1410–1419.

[2] Paschen, U., Pitt, C., & Kietzmann, J. (2020). Artificial intelligence: Building blocks and an innovation typology. *Business Horizons*, 63(2), 147–155.

[3] Paschen, J., Wilson, M., & Ferreira, J. J. (2020). Collaborative intelligence: How human and artificial intelligence create value along the B2B sales funnel. *Business Horizons*, 63(3), 403–414.

[4] Leone, D., Schiavone, F., Appio, F. P., & Chiao, B. (2021). How does artificial intelligence enable and enhance value co-creation in industrial markets? An exploratory case study in the healthcare ecosystem. *Journal of Business Research*, 129, 849–859.

[5] Lin, W. L., Yip, N., Ho, J. A., & Sambasivan, M. (2020). The adoption of technological innovations in a B2B context and its impact on firm performance: An ethical leadership perspective. *Industrial Marketing Management*, 89, 61–71.

[6] Murgai, A. (2018). Transforming digital marketing with artificial intelligence. *International Journal of Latest Technology in Engineering, Management & Applied Science*, 7(4), 259–262.

[7] Farrokhi, A., Shirazi, F., Hajli, N., & Tajvidi, M. (2020). Using artificial intelligence to detect crisis related to events: Decision making in B2B by artificial intelligence. *Industrial Marketing Management*, 91, 257–273.

[8] Lau, R. Y. (2007). Towards a web services and intelligent agents-based negotiation system for B2B eCommerce. *Electronic Commerce Research and Applications*, 6(3), 260–273.

[9] Prior, D. D., & Keränen, J. (2020). Revisiting contemporary issues in B2B marketing: It's not just about artificial intelligence. *Australasian Marketing Journal (AMJ)*, 28(2), 83–89.

[10] Huang, M. H., & Rust, R. T. (2018). Artificial intelligence in service. *Journal of Service Research*, 21(2), 155–172.
[11] Wilson, R. D., & Bettis-Outland, H. (2019). Can artificial neural network models be used to improve the analysis of B2B marketing research data? *Journal of Business & Industrial Marketing*, 35, 495–507.
[12] Blake, M. B. (2002, July). B2B electronic commerce: Where do agents fit in? In *Proceedings of the AAAI-2002 Workshop on Agent Technologies for B2B E-Commerce*, Edmonton, Alberta, Canada.
[13] Martínez-López, F. J., & Casillas, J. (2013). Artificial intelligence-based systems applied in industrial marketing: An historical overview, current and future insights. *Industrial Marketing Management*, 42(4), 489–495.
[14] Tahvola, K. (2020). Leveraging artificial intelligence in B2B Markets.
[15] Paschen, J., Kietzmann, J., & Kietzmann, T. C. (2020). Unpacking artificial intelligence—How the building blocks of artificial intelligence (AI) contribute to creating market knowledge from big data.
[16] Chen, A. P. S., Chansilp, K., Kerdprasop, K., Chuaybamroong, P., Kerdprasop, N., Shun-Fa, H., . . . Kaium, M. A. (2020). B2B marketing crafts intelligence commerce: How a Chatbot is designed for the Taiwan Agriculture Service. *International Journal of e-Education, e-Business, e-Management and e-Learning*, 10(2), 114–124.
[17] Vladimirovich, K. M. (2020). Future marketing in B2B segment: Integrating artificial intelligence into sales management. *International Journal of Innovative Technologies in Economy*, 4(31).

Chapter 10

A Systematic Literature Review of the Impact of Artificial Intelligence on Customer Experience

M. A. Sikandar, Praveen Kumar Munari, and Meghraj Arli

Contents

DOI: 10.4324/9781003206316-10

10.1 Introduction

Artificial intelligence (AI) is the most significant technological advance; these machine-learnt technologies benefit both businesses and customers. The journey that a customer takes from pre-purchase to post-purchase can be defined as the user's experience (Braun and Carriga, 2017).

Personalization and AI interaction have the potential to bring customers closer to online retailers (Parkes, 2018). AI is meant to improve customer experience and thus help businesses adapt to "servitization."

However, machine learning (ML) does not evolve changes in the path when online retailing is done (Amazon Annual Report, 2018). It also influences how customers in the online retailing shop in traditional brick-and-mortar stores. As a result, the gap between offline and online retailing (brick-and-mortar stores) is closing (e-commerce) (Silver, 2016).

Department stores, as well as internet sites in the omnipresent world, could not be regarded as independent bodies. Instead, the two are advantageous to one another. As a result of this omnichannel view of commerce, customers' purchase paths are becoming much more complicated.

According to Hogg (2018) in an article written for Google, a travel guide with exits, research, and conversation across the path is identical to a customer lifecycle, whenever it is necessary to persuade, customers to choose and stay with your company instead of turn to a rival (LG, 2019). As a result, it is difficult for marketers to create a value offer that is tailored to the customer's unpredictable desires. As a result, businesses must plot their customers' paths to purchase by using their leverage technology advancements to optimize consumer loyalty in the purchasing decision (Kaci, Patel, and Prince, 2014). Machine learning (ML) can increase user travel; however, companies must know how these innovations affect the customer perspective in this ever-changing environment

10.2 Conceptual Backgrounds: Literature Review

10.2.1 Defining Experience

Customer experience refers to a user's intrinsic and impartial reaction to interacting directly with a company. Even during purchase, use, and support of a product, the customer usually initiates direct communication and experience. A series of encounters between an individual and a product, a business, or a feature of its organization that conveys a reflex is referred to as consumer experience. This is a highly personal observation that necessitates the customer's involvement on many levels. Its assessment is based on a comparison of the needs of a customer and the sensations obtained from experiences with the business and its offerings in relation to various multiple touchpoints or instances of contact (Bolton, Gustafsson, McColl-Kennedy, Sirianni, and Tse, 2014).

The aforementioned concept of customer engagement is multidimensional, encompassing sensory, cognitive, adaptive, physical, and emotional elements (DePillis and Sherman, 2019). Secondly, it understands the importance of both logical and emotional dimensions of customer service.

10.2.2 Customer Experience Management Definition

In a Seminal Article published in 1998, Joseph Pine II and James H. Gilmore coined the word "Customer Experience Management." The writers stressed the importance of making a lasting impression on the consumer (Bhandari, Rama, Seth, Niranjan, Chitalia, and Berg, 2017).

Customer experience is a clear reflection and flawless implementation of the emotional link and interactions you intend your consumers should have with any brand through delivery channels and contact points. Later, customer interaction management is understood as the method of efficiently managing a buyer's entire experience with a brand or service, according to each (Lutz, 2017). The significance of incorporating different kinds of customer satisfaction through multiple touchpoints was emphasized.

10.2.3 Artificial Intelligence

Crittenden, Biel, and Lovely (2019) define AI as the concept of machines capable of performing activities intelligently; delivering useful, efficient solutions to issues. Individuals also assert that AI is related to other technical advancements including machines and data science. Computer learners have the opportunity of computers to learn on their own without the use of comprehensive programming, and robots are capable of performing tasks that humans would normally perform (Amazon Echo, 2019; Amazon Echo Auto, 2019; Amazon Locker, 2019).

Geisel (2018) defines AI as follows: a truly artificially intelligent system by itself will read. We speak of ML, which could shape links while depending on pre-programmed conduct equations and extract meanings. True AI can increase its strengths or experience through earlier installments, becoming wiser and much more aware as it progresses (Levy, 2018).

AI may be considered to be "intelligent" and "automated" systems. These meanings do not seem quite precise and are generally very technical. In the course of this thesis, to provide a better picture of the implications of AI, I will explore different IC strategies and developments in sales and economics (Chamatkar and Butey, 2015).

10.2.4 Customer Personalization

It is important to recognize the actions of customers, as per Chatterjee (2018) because that knowledge would be used to define and address a targeted audience adequately in line with your wants and requirements. The importance of being is emphasized by Google (2015) applying the principles of "being rapid." One of the measures organizations could undertake when doing operations is to anticipate needs. Firms should look at previous actions of a customer, can classify them, and include a relevant message tailored to the individual. It is important to consider consumer conduct and provide the consumer with an unfriendly experience. Chatterjee (2018) discusses a "Consumer Behaviors DNA" algorithm. This algorithm (which employs pattern mining) assists businesses in identifying various types of customer behavior.

Companies can use this data to better segment customers and gain a better understanding of their needs/preferences. With this understanding and information about purchasing behavior, businesses can use recommender systems to better target their customers and personalize their services.

10.3 Customer Experience and Journey

Customer experience is just how consumers view the organization experiences. A good approach from either the customer's perspective must be productive, helpful, and fun. The particular aspect of the customer perception, a client's character, restricts its management utility for forecasting and controlling reasons at a specific moment and place in a particular case. Many enterprises consider customer satisfaction (CEM) as a customer relationship management successor (CRM). But interventional integration problems are becoming even more challenging.

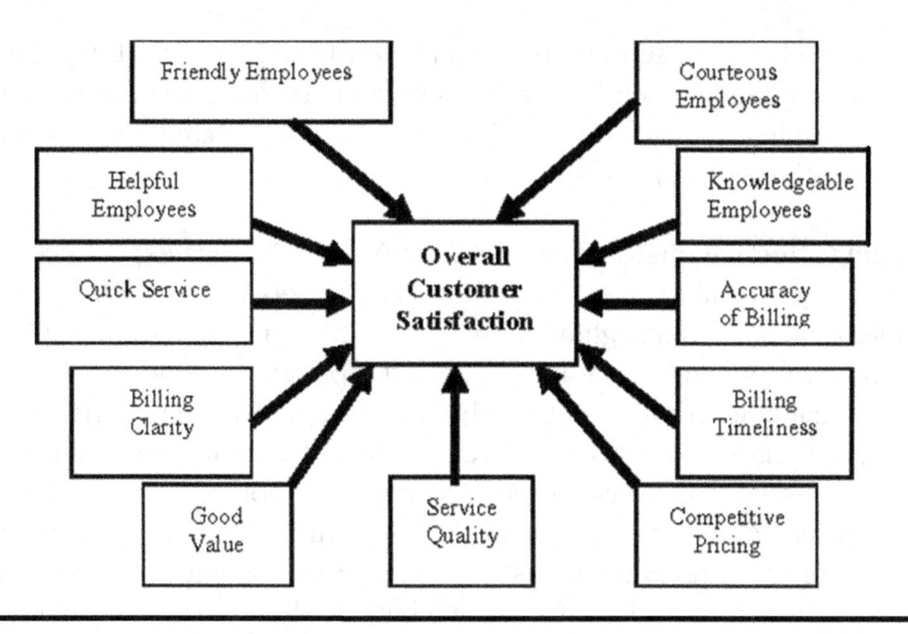

Figure 10.1 Factors that affect customer satisfaction.

For this analysis, five main criteria for measuring customer service were considered in terms of service quality and service retention, according to UCTI, the University College of Technology and Innovation of Asia Pacific. With the aid of Figure 10.1, we can quickly obtain a general summary of customer satisfaction.

A customer journey map provides you with more information about the consumer, allowing you to go beyond what you already know. Many companies regard the customer journey as something observable – the point at which the client engages with the company. However, this is not the case, and only accounts for a small portion of the whole consumer experience. Making a customer journey map forces you to consider the components of the trip that you don't see but are as important to the overall experience.

10.3.1 Types of Consumer Experiences

By the study of a few research papers like Braun and Carriga's (2017) philosophical analyses, Schmitt (2011) stated that five kinds of relations like hear, feel, perspective, act then relate.

Sense: It's a sensory value that offers an amazing event stimulus by the clear response to the five human senses: consumer sense, hearing sense, consumer feel, tasting sensation, and odor sensing. On the road, for example, a Jaguar gives a case of a car responsive experience, this attraction is in sensual meaning. Design sense. A Ferrari, on the other hand, has a responsive and exciting perceptual value.

i. **Feel (affective customer experience):** The affective experience calls on the personal feelings and ideas of consumers to create affective interactions from a slightly optimistic point of view market behaviors (e.g. for non-competitive moods, brand, operation, or industrial non-sustainable food product) to high pleasure and pride emotions (e.g. sustainable market, technology). Feels as though it is, attributes of feeling that are related to consumer sentiments and emotions. There are rules of emotional experience that we have to drink a cup of coffee store, the excitement we're living cherishing a trip to Paul Krugger's National Park, and there are loose feelings that we have to drink a cup of coffee store.
ii. **Consider the following (cognitive/creative business performance):** I believe in the mind intending to develop cognitive interactions that produce outcomes while also involving our clients' innovative thinking. Think; attempts by suspense, excitement, and agitation to involve consumers' inductive and deductive thought. The latest technological merchandise is thought of as campaigns. Think of artistic knowledge principles that attract people reasoning abilities via the life experience that resolve issues and cognitive problems (Chamatkar and Butey, 2015).
iii. **Physical customer service (Act):** This Act is designed to influence physical perceptions and lifestyles. Marketing enhances the lives of consumers by improving physical environments, alternative practices and relations, as well as new ways of doing things. Ethical methods to personality development associated with the act are usually motivators, inspiring, and creative by positive examples, such as celebrities or well-known players (Chamatkar and Butey, 2015). Act is about standards of conduct that relate to resistance training, attitudes, and interpersonal interactions.
iv. **Relate (social-identity customer experience):** RELATE includes elements of the selling of senses, feelings, thought, and actions. That being said, relationships extend far beyond intimate, secret thoughts of the partner and therefore connect to "personal experiences" and connect the person to his/her values, other persons, or community (Chamatkar and Butey, 2015). The connection is about relative values of experience which call for personal liberalization. Nike's affection means that

a customer puts on the rear of the wrist a tattoo to display him/her the company.

10.3.2 Artificial Intelligence and Customer Experiences

The ability of AI by simplifying consumer experiences and purchasing trips is to enhance the importance in the business world. Marketing companies and intelligence experts have to cooperate but not fight to achieve this (Chamatkar and Butey, 2015). Salespeople should rethink their service delivery and concentrate on AI to improve confidence or pace in operations, improve the volume and make better decisions, and create more efficient customization. When people and sophisticated technology partner, companies can see their greatest cost savings. Companies need workers to develop technology and clarify their actions to ensure the efficient use of processors (Freitas, Gastaud Maçada, Brinkhues, and Zimmermann, 2016).

10.4 Research Methodology

There hasn't been a lot of research done on the subject of AI in management. Although AI is also a relatively young field, the majority of research in this field is either continuing or unfinished. However, there is AI research available in other fields that contributes to our understanding of AI in the industry.

The research I conducted for this thesis was in the field of marketing, specifically the consumer experience and journey. To evaluate the influence of AI on the customer journey, we conducted a literature review. This analysis takes into account previous research on the topic. The objective of this research work is to allow the researcher to understand the theory by research questions. A lot of information available inquiries in an internet source are part of the research quest. Keywords used in publications contain (a) customer experience; (b) e-commerce (Piyush, Choudhury, and Kumar, 2016) and (c) AI.

The following were the conditions for inclusion: English language; business, management, and accounting and to reduce the number of results, a separate search was conducted for more general knowledge on AI and the user path, using different inclusion criteria such as region and time. In addition, I used Amazon.com as a case study and analyzed it using blogs, press releases, and news stories. Companies should use Amazon as a best practice example for improving their customers' experiences with technology like AI, according to this case study (Viktor, 2019).

10.5 Results and Conclusion

Frameworks to demonstrate the effect of classification models and chatbots can have on service quality, as well as to map such deep learning techniques through the customer journey, as previously discussed. The activities of customer experiences are as shown in Figure 10.2.

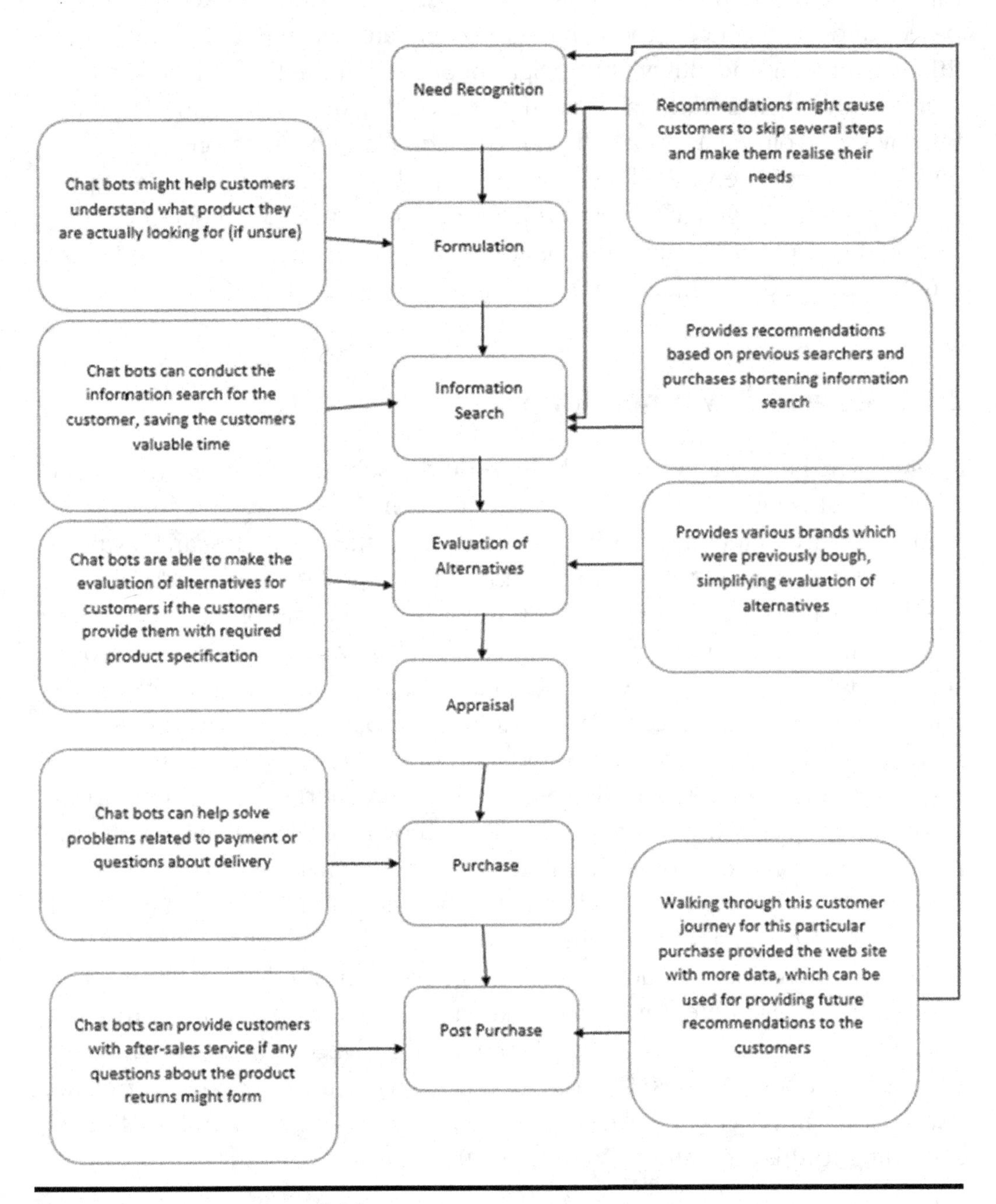

Figure 10.2 Customer activities.

However, understanding the consumers' behavior and desires are critical to do so. Consumer experience is affected by both practical and psychological factors, and it is important to gain a better understanding of customer behavior by determining how AI techniques affect these. Customer personalization is possible with recommender systems, while customer experience can be improved with chatbots. The following is how the customer experience is defined: The customer experience is made up of the interpretations that buyers have while communicating with a business.

Customers' qualitative reactions would be favorably affected if AI could be used to improve interactions at different touchpoints. Many of the better experiences from the enhanced shopping journey are brought together, from the beginning to the end of the transaction.

The impact of AI on the consumer experience during the shopping experience was investigated in this report. This has been discussed as part of a broader context focused on previous research. With the increasing significance of providing customers with a special experience, literature that illustrates the impact of technological advancements on this experience is required. Piyush, Choudhury, and Kumar (2016) suggest a structure for new value development channels in the digital era. Automation, individualization, ambient embeddedness, interaction, openness, and control are some of these outlets.

References

Amazon Annual Report. (2018). Amazon Annual Report 2018. Retrieved on 02–06–2019 from https://ir.aboutamazon.com/static-files/0f9e36b1-7e1e-4b52-be17-145dc9d8b5ec

Amazon Echo. (2019). Amazon Echo. Retrieved on 02–06–2019 from www.amazon.com/all-new-amazon-echo-speaker-with-wifi-alexa-dark-charcoal/dp/B06XCM9LJ4

Amazon Echo Auto. (2019). Amazon Echo Auto. Retrieved on 02–06–2019 from www.amazon.com/Introducing-Echo-Auto-first-your/dp/B0753K4CWG

Amazon Locker. (2019). Amazon Locker. Retrieved on 01–06–2019 from www.amazon.com/b/ref=amb_link_366591722_2?_encoding=UTF8&node=6442600011&tag=bisafetynet2-20

Bhandari, A., Rama, K., Seth, N., Niranjan, N., Chitalia, P., Berg, S. (2017). Toward an Efficient Method of Modelling "Next Best Action" for Digital Buyer's Journey in B2B. In: *International Conference on Machine Learning and Data Mining in Pattern Recognition*, MLDM 2017, Lecture Notes in Computer Science, 10358, Springer, Cham, pp. 107–116.

Bolton, R.N., Gustafsson, A., McColl-Kennedy, J., Sirianni, N.J., Tse, D.K. (2014). Small Details That Make Big Differences: A Radical Approach to Consumption Experience as a Firm's Differentiating Strategy. *Journal of Service Management*, 25(2), pp. 253–274.

Braun, A., Carriga, G. (2017). Consumer Journey Analytics in the Context of Data Privacy and Ethics. *Digital Marketplaces Unleashed*, pp. 663–674, Springer Nature Switzerland.

Chamatkar, A.J., Butey, P.K. (2015). Implementation of Different Data Mining Algorithms with Neural Network. In: *2015 International Conference on Computing Communication Control and Automation*, 26–27 Feb., Pune, India, IEEE, pp. 374–378.

Chatterjee, R. (2018). Amazon's 2 Most Powerful Rivals Just Decided to Team Up. *Business Insider*. Retrieved on 02–06–2019 from www.businessinsider.com/r-walmart-microsoft-in-partnership-to-use-cloud-tech-2018-7?international=true&r=US&IR=T

Crittenden, W.F., Biel, I.K., Lovely, W.A. (2018). Embracing Digitalization: Student Learning and New Technologies. *Journal of Marketing Education*, 21(1), pp. 5–14.

DePillis, L., Sherman, I. (2019). Amazon's Extraordinary Evolution: A Timeline. *CNN*. Retrieved on 23–05–2019 from https://edition.cnn.com/interactive/2018/10/business/amazon-history-timeline/index.html

Freitas, J.C., Gastaud Maçada, A.C., Brinkhues, R., Zimmermann, G. (2016). Digital Capabilities as Driver to Digital Business Performance. In: *22nd Americas Conference on Information Systems*, San Diego, USA, pp. 1–5.

Geisel, A. (2018). The Current and Future Impact of Artificial Intelligence on Business. *International Journal of Scientific & Technology Research*, 7(5), pp. 116–122.

Google. (2015). Micro-Moments: Your Guide to Winning the Shift to Mobile. *Think with Google*. Retrieved on 28–06–2019 from www.thinkwithgoogle.com/marketing-resources/micro-moments/micromoments-guide-pdf-download/

Hogg, S. (2018). Customer Journey Mapping: The Path to Loyalty. *Think with Google*. Retrieved on 28–06–2019 from www.thinkwithgoogle.com/marketing-resources/experience-design/customer-journey-mapping/

Kaci, S., Patel, N., Prince, V. (2014). From NL Preference Expressions to Comparative Preference Statements: A Preliminary Study in Eliciting Preferences for Customised Decision Support.

Levy, S. (2018). Inside Amazon's Artificial Intelligence Flywheel. *Wired*. Retrieved on 02–06–2019 from www.wired.com/story/amazon-artificial-intelligence-flywheel/

LG. (2019). LG Fridge Alexa Amazon. Retrieved on 02–06–2019 from www.lg.com/us/refrigerators/lg-LNXS30996D-door-in-door

Lutz, A. (2017). Amazon Is Officially Buying Whole Foods—Here's Everything That Will Change for Customers. *Business Insider*. Retrieved on 23–05–2019 from www.businessinsider.com/amazon-buys-whole-foods-changes-2017-8?international=true&r=US&IR=T

Parkes, T. (2018). How One Retailer Is Shifting to an AI-First Mentality. *Think with Google*. Retrieved on 28–06–2019 from www.thinkwithgoogle.com/marketing-resources/experience-design/1-800-flowers-voice-assistants/

Piyush, N., Choudhury, T., Kumar, P. (2016). Conversational Commerce a New Era of E-Business. In: *2016 International Conference System Modelling & Advancement in Research*.

Schmitt, B. (2011). Experience Marketing: Concepts, Frameworks and Consumer Insights. Foundations and Trends in Marketing, 5(2), pp. 55–112.

Silver, C. (2016). Amazon Announces No-Line Retail Shopping Experience with Amazon Go. *Forbes*. Retrieved on 23–05–2019 from www.forbes.com/sites/curtissilver/2016/12/05/amazon-announces-no-line-retail-shopping-experience-with-amazon-go/#f7cf2d12326e

Viktor, R. (2019). Using Adversarial Training to Recognize Speakers' Emotions. *Amazon Alexa Blogs*. Retrieved on 22–05–2019 from https://developer.amazon.com/blogs/alexa/post/2d8c2128-eec9–44cc-9274-444940eb0a4d/using-adversarial-training-to-recognize-speakers-emotions

Chapter 11

The Impact of Artificial Intelligence on Customer Experience and the Purchasing Process

Laxmi Shaw, Megha Mankal, and Chinnapani Kiran Kumar

Contents

DOI: 10.4324/9781003206316-11

11.1 Introduction

The concept mainly focusses on implementing artificial intelligence in various fields especially in business to customer segment to enhance the customer experience. Artificial intelligence (AI) plays a significant role for customer's engagement and customer's buying journey.

The role of AI throughout customer's buying journey is at different phases like stage of awareness, phase of consideration, process of purchasing, and support phase. Customer experience can be explained with two important dimensions namely personalized customer service and after-sales customer support. Due to globalization and industrial revolution, fulfilling customer expectations which are at higher level has become challenging task for all industrial segments [1].

AI has gone through various innovations by introducing various platforms like augmented reality, virtual reality and mixed reality. The mentioned platforms are having wide applications especially in customer's buying journey. It will respond and provide services to customer based on demographics and lifestyle [2].

Chatbots are almost utilized as artificial salesperson during the purchasing process, by providing appropriate information of products to the customer. Customer purchase journey geos through three major important stages of transactions namely pre-transaction, transaction, and post-transaction [2].

The aim of our review is to identify the factors that have greater influence on customer experience, customer's buying journey using various innovations, and applications of AI.

In the coming discussions, we elaborate theoretical aspects of customer experience and customer buying journey. Also, we would like to discuss on definitions and concepts of applications and innovations of AI and thereafter to discuss a separate analysis for both customer experience and consumer buying journey using applications of AI.

11.2 The Customer Experience: A Literature Review

The significance of customer experience and customer buying journey is well cited in the literature. Also, how AI influences the customer buying journey and resulting to achieve greater customer loyalty and profit is well explained [1].

The "digital revolution" has fundamentally changed the consumer experience over the last 20 years. In his book Becoming Digital, Nicholas Negroponte (1995) [3] described the transition from "atoms" to "bits" to explain digital technologies and their benefits [2].

11.2.1 Concepts and Definitions

Customer experience can be explained in various ways of approach; among them, other prominent customers typically like friends and family are also causes for differences in one's own customer experience. The research also identified there are different customer factors discussed in literature. The prominent factors include social factors, task factors, time factors, product involvement, and financial resources [4].

The literature proposes definitions for customer experience dimensions such as customer service and after-sales customer care [1].

The concepts of augmented reality, virtual reality, mixed reality are explained well in the discussion. And the applications like virtual assistants, chatbots, and robots are well identified and explained in the discussion [2].

Drawing from those concepts and definitions, we suggest defining customer experience as "[t]he experience which physically and emotionally connected with customer throughout the customer buying journey across various phases of purchase process."

11.2.2 The Dimensions Included in Customer Experience

In addition to the two dimensions of customer experience mentioned earlier namely customer service and after-sales customer support, few other multi-dimensions for customer experience are identified.

The major dimensions of customer experience identified are SERVQUAL model, eTailQ Experience, E-S-QUAL, consumption experience model, customer experience model, conceptual model of customer experience, and finally, EXQ framework [5].

All the above dimension models of customer experience have sub-dimensions within each model. The popular model named SERVQUAL consists of factors like reliability, responsiveness, assurance, empathy and tangibility. The sub-parts of the eTailQ experience include web design, fulfillment/authenticity, security/seclusion, and real-time customer support [5].

Performance, fulfillment, device availability, and privacy are all sub-dimensions of the E-S-QUAL model. Sensual/intuitive, non-cognitive, physical/behavioral, social, and cognitive variables all play a role in the consumption experience model. Sensorial, sentimental, rational, realistic, standard of living, and associative components all together build up the customer experience model [5].

Another level of complexity of the social climate, interface of the service, retail atmosphere, variety, price, interactions of consumers in alternative channels, and retail brand are all parts of the customer experience conceptual model. Product experience, outcome concentration, moments of reality, and peace of mind make up the final component of the EXQ system [5].

11.2.3 The Customer Experience Measurement

Although theorizing about customer experience has progressed significantly, measuring customer experience in service research has received little attention (Lipkin, 2016). In service analysis, the terms customer experience and service experience are often used interchangeably, but they are not synonymous (Klaus and Maklan, 2012) [6].

To evaluate customer experience in retail sector, two approaches are suggested. First approach measures customer experience based on customer's perceptions or evaluative judgements of elements leading to service experience. Second approach measures customer experience based on structural equation modeling developed [6].

Machine learning (ML) algorithm is an important tool to understand the customer behavior. During buying process, customer might come across various advertisements and promotions for which identifying perceptions and human behavior is sometimes more complex to quantify. Sometimes, we couldn't fully understand how the algorithm works for these kinds of scenarios [7].

11.3 Artificial Intelligence in the Public Sector

The potential applications of AI are vastly increasing with day-to-day requirements of consumers involved across the globe. Among them, public sector benefits from AI through several ways like public safety, social welfare, and public health [1].

Using AI in public safety helps in reducing time delays of patients being admitted. Also, prioritizing patients who need ambulance and who needed to treat on site [1].

Using AI in social welfare mainly advises to public employees to resolve complex issues arise frequently. And finally using AI in public health is to identify the patients having similar symptoms in different locations and spread awareness of the disease [1].

When the three aspects are applied to the public sector, it is possible to provide better service to all users while still increasing national security and satisfaction.

11.4 Artificial Intelligence in Retail Sector

The retail industry is undergoing a significant transformation. A business as a whole is attempting to keep up with rapidly changing consumer data and to have sufficient value to transform to online mode rather than traditional mode [8].

In retail sector, AI-controlled brilliant mirror utilized stores for making virtual perception of garments to the customer as an experience without wearing and letting how it looks bringing new age environment throughout [8].

The key technologies involved for above mentioned smart mirrors have RFID controlled garment racks, gyro-sensors, and function low vitality chips. These mirrors are also useful for beauticians to change lighting of the mirror as per the mood set of customers and make customer delight while providing services accordingly. These technologies will enhance for future visit of customers to the store [8].

11.5 Artificial Intelligence in B2B E-commerce

The growth of e-commerce has increased the value of SCM as businesses reengineer processes as they migrate to the internet.

A range of AI-based SCM problem-solving methods are available. Most of them are agent-based, with each agent in charge of one or more SCM operations. Most agent-based e-commerce approaches concentrate on transactional information, that is, the knowledge needed to negotiate on price, delivery date, product quantity, and so on [9].

Some agent-based methods will deal with knowledge that isn't transactional, such as SCM knowledge. Humans or organizations can be modeled using certain agent-based approaches to solving SCM problems. The models are used for risk value analysis and simulation (Swaminathan et al. 1998) [9].

The foremost goal of AI is to build and create a system that is capable of imitating the human behavior. An e-commerce system with AI components can function more "naturally" to its users [9].

Key dimensions that play a major role in deciding and evaluating the AI applications to e-commerce include naturalness, performance, and creating a better user interface that facilitates interactions [9, 10].

11.6 Challenges of Artificial Intelligence

Even with better enhancements of AI happening day to day, but still they are situations where machines or processes going out of control leading to effect human race. These constitute challenges of AI in customer experience and customer buying journey. The major challenges identified in the discussion with examples are as follows.

Microsoft used AI to build a bot that can have automated conversations with Twitter users while mimicking their language. The bot has become offensive and has been labeled racist in less than 24 hours [1].

Google has apologized for a bug in Google Photos that caused images of black people to be mislabeled as gorillas. A sizable portion of customers believe that using AI in investments is riskier than it is advantageous [1].

When it comes to the complexities of AI in marketing, the most challenging part of AI integration is technological compatibility. Person named Waqar Haider says the company is successful in integrating their system with major CRM systems to solve the issue of compatibility. It remains a significant obstacle for us, and the organization is constantly working to strengthen the operation [11].

Coming to the challenges for AI in healthcare, there are various factors identified. The main challenges include the need for specific architecture in businesses, public perceptions of AI, the need for privacy and data security, and the need for high reliability and service quality [12].

11.7 Benefits of a Satisfied Customer

Customer satisfaction is important because it encourages consumers to feel a sense of togetherness, sentimental connection, and allegiance of the brand. Customer satisfaction is defined as "whether a product or service meets a customer's needs or demands" or "whether a customer believes he receives benefits he always wanted from the goods or services for which the amount has been paid to a specified firm." [13].

According to a survey that is held, the following observations are drawn:

> 86% of adults spend more on the product of a particular brand they feel loyal about. If their shopping experience is positive, more than 70% of people tell other shoppers about it. 43% of the adults purchase so called inferior products if their experience with the product is positive [1].
>
> The above three findings demonstrate that providing a superior customer experience is critical for increasing revenue, customer loyalty, and customer base. Delivering superior service creates a competitive edge [1].

11.8 Research Methodology

Regarding the methodology, a qualitative study on Palestine-based companies along with CEO and IT specialists are chosen for the analysis and through that later quantitative findings are identified in the discussion [1].

For analyzing consumer buying journey and purchase process, applications of AI like chatbots are induced to know the levels of performance in purchase process [14].

11.8.1 Population

Based on intense literature review, as mentioned earlier, technology-based companies are considered where CEO and IT specialists along with internet uses residing in Palestine West Bank are taken as population for the research [1].

11.8.2 Sample Size

Two companies are considered for investigation whether they can go for AI implementation [1].

11.8.3 Data Sources

Primary Data: Firsthand data collected based on research.

Interviews: Interview with CTO of online booking company named "YAMSAFER." Interview with a CEO of Palestinian banking company names The National Bank (TNB) [1].
Questionnaire: The closed-ended questionnaire was distributed to investigate internet users' actions on AI [1].
Secondary data: Obtaining data that has already been analyzed by other researchers [1].

Now for analyzing consumer buying journey using chatbots, Krippendorff (2004) [15] is a distinct technique used when researchers try to avoid receptive situations on two grounds: misrepresenting the findings, undermining the study's validity, and data manipulation by the thesis or origins being studied [14].

Information is obtained through systematic and face-to-face observation, wherein ten retailers are being observed conducting their retailing operations in the Romanian online market in the position of a "mysterious customer." [14].

The content review concentrated on Romanian retailers with the highest number of consumers. Because of the growth of electronic commerce and the potential for multiplying e-commerce transactions established in accordance with the Western model, we have considered the Romanian market for the analysis [14].

The websites were selected based on the number of people who visited online stores in August 2020. The criteria for selection are supported with a fact that nearly 70% of Romania's 199.4 lakh citizens spend time surfing the internet.

11.8.4 Conceptual Model

Figure 11.1 depicts the related variables in this analysis that the authors used to evaluate the study hypothesis to emphasize the relationship between AI and customer experience. The independent variable is AI, whereas the dependent variable is customer experience, which is divided into two dimensions: personalized customer service and after-sales customer support.

The consumer purchasing journey model will show how simple chatbots are used, their consistency and effectiveness, response time, and the value of customer responses. The information was coded (Table 11.1) so that the results could be analyzed consistently.

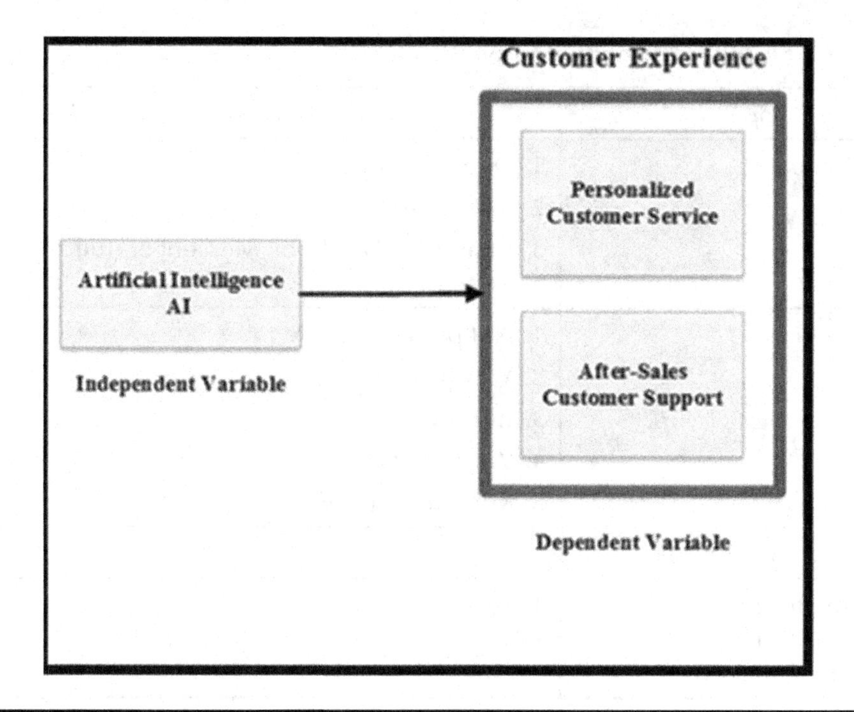

Figure 11.1 Study hypothesis.

To interpret the data, the results of horizontal analysis (to calculate the number of occurrences of the code) and vertical analysis are being used.

11.9 Discussion of the Findings

For research hypotheses in this section, as well as the corresponding discussion that goes among each hypothesis, both qualitative and quantitative data findings are included in this section.

H0–1:There exists no significant relationship between AI and customer experience.
H0–2: There exists no significant influence of AI on customer experience.

According to Table 11.2, the authors of the research paper applied correlation to analyze the significance of relationship between AI and customer experience; findings indicate that the two variables have a direct and moderately positive relationship [1].

Table 11.1 Responses Coded in Content Analysis

Topic	*Coding*
Using Chatbots	1. Dont use any chatbots 2. Instant replies 3. Chatbot in Facebook Messenger (full set version)
Response time	1. Not publicly visible 2. Very prompt 3. In a few hours 4. Within a day 5. More than a day.
Instant message Quality	
– Information on personalized recommendations	1. Yes 2. No
– Product cost information	1. Yes 2. No
– Information on personalized assistance	1. Yes 2. No
– Information on popular personalized products	1. Yes 2. No
– Information on the availability of products in stock	1. Yes 2. No
– Using a custom addressing formula	1. First name only 2. Last name only 3. First name and last name 4. No first name, no last name
Relevance of replies	1. High 2. Medium 3. Low
Performance	1. Basic sending of automatic messages 2. Proactively recommend content

Table 11.2 Coefficient of Correlation between Artificial Intelligence and Customer Experience

Dimensions	1	2
Artificial Intelligence	1	0.514**
Customer Experience	0.514**	1

Table 11.3 ANOVA Table for the Econometric Model of Customer Experience

Model	*Sum of squares*	*df*	*Mean square*	*f*	*Sig.*	*R*	*R-Square*
Regression	13.571	1	13.571	22.669	0.000		
Residual	18.832	89	0.306			0.514	0.264
Total	32.403	90					

So, hypothesis (H0–1) concludes that there exists a significant relationship between AI and customer experience.

R-square and ANOVA tests are employed to identify the results relating to the hypothesis; Table 11.3 shows that the F-distribution of 1 and 89 df has a significant value of 22.669. The significance of the regression is established by using the F-test, with a value of p less than 0.05; hence, the results suggest that AI and customer experience have a significant relationship [1].

With the help of R-square and ANOVA reports, it has been identified that there exists a positive relationship between AI and customer experience. As $R^2 = 0.264$, AI estimates 26.4% of the difference in customer experience [1].

The effect of AI on the customer experience was explained by the authors in form of two aspects (personalized customer service and post-sales customer support), and the R-square and ANOVA summary statistics are mentioned in the above table [1].

It has been identified that AI accounts for 22.9% of the variance in personalized customer service mentioned in Table 11.4, whereas after sales customer support accounts for 7% in Table 11.5 [1].

Now moving on to the results and discussion of consumer buying journey using chatbots, the following content analysis table is explained.

The results of the study paint a picture of support services in the pre-purchase stage of the customer's journey, where the customer intends to connect with the provider by employing Facebook Messenger. By using

Table 11.4 ANOVA Table for an Econometric Model of Personalized Customer Experience

Model	*Sum of Squares*	*df*	*Mean Square*	*f*	*Sig.*	*R*	*R-Square*
Regression	9.93	1	9.93	24.669	0.000		
Residual	33.409	89	0.403			0.479	0.229
Total	43.420	90					

Table 11.5 Post-Sales Customer Support (ANOVA TABLE)

Model	*Sum of Squares*	*df*	*Mean Square*	*f*	*Sig.*	*R*	*R-Square*
Regression	3.641	1	3.641	6.294	0.014		
Residual	48.014	89	0.578			0.265	0.07
Total	51.781	90					

horizontal and vertical analyses, it has been found that 100% of the major retailers (Table 11.6) are not using the entire full version of the chatbot [14].

The effect of chatbot content on final customer purchasing behavior was explored by looking at the quality aspects from the perception of the communication started. The following description of each group integrated in the content analysis was used to present the findings as concisely as possible.

Although no complete version of a chatbot has been introduced in any of the ten online stores studied (Figure 11.2), the existence of a particular form of chatbot in all ten cases analyzed defines the scenario. This is reflected in the automatic alerts that retailers have set up, as well as the interest in using AI technology to initiate online experiences with consumers [14].

The supply of in-stock items, the price of customizable products, specialized common products, personalized reviews, personalized assistance, and details on the use of the first name were all considered while evaluating the quality of the chatbot [14].

The content analysis coded the performance in two ways: "sending simple messages" and "recommending content." The condition in which material is suggested to the user was not encountered in any of the ten cases examined. Using the related "simple chatbots," all of the prominent retailers sent general conversations, which defined the results obtained [14].

Table 11.6 Online Store Ranking (Romania)

Place	*Website*	*Users*	*Visits*	*Views*
1	www.pretzmic.ro	51,904	72,308	269,250
2	www.baterii.lux.ro	43,128	59,857	158,099
3	www.uscatorrufe.ro	25,992	31,429	67,343
4	www.wainertools.ro	25,958	31,763	69,993
5	www.magazin-unelte.ro	19,658	24,002	90,991
6	www.vintagetime.ro	19,407	23,954	64,382
7	www.parfumas.ro	12,367	25,350	248,482
3	www.fabricdaelenjerii.ro	11,874	18,720	76,668
9	www.lexservice.ro	11,774	15,343	86,643
10	www.gamestore.ro	11,259	16,279	65,688

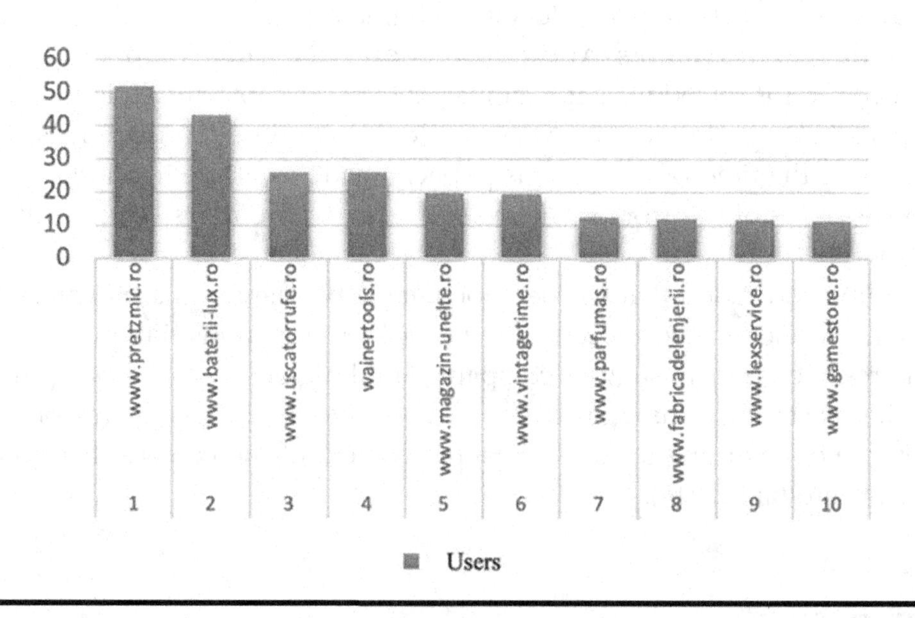

Figure 11.2 Online store ranking (Romania) based on the number of consumers.

11.10 Conclusion and Future Work

Using theoretical and conceptual context, the influence on customer experience and consumer buying journey is explained with separated analysis identified. The influence on customer experience and future work is explained as follows.

The results of the correlational and regression studies show that AI and customer experience have a good relationship. AI and delivering customized customer service and after-sales customer care are inextricably linked [1].

Personalized customer service has a positive effect on customer experience throughout the buying journey of the customer. Customers can wait less and have a better experience through the use of AI in customer contact centers and in other areas of service [1].

The study recommended to enhance the customer experience in the awareness stage. It also advised companies to offer personalized to the customer to enhance their experience. According to the report, AI can be used in call centers and other after-sales support facilities to reduce customer wait times [1].

Now, the influence of AI tool chatbots on consumer buying journey and recommendations are explained as follows.

Consumer frustration is fuelled by the poor quality of content displayed by simple chatbots, particularly during the pre-purchase stage, when users want to communicate with retailers in real time [14].

According to the authors, AI can help boost personalized support features, even if it is introduced at a simple level, if chatbots are designed to be capable of engaging in a conversation for a long time on the request of users, and if the time of response is publicly made available in all of the Facebook pages of the store retailers, as a result where users loyalty will magnify [14].

It can be concluded that the decisions made by managers are expressed in actions developed by retailers in their business functions [14].

The research finders suggest companies in the global world to keep track of the levels at which the target public is at and to assess the degree of acceptance of emerging technology in the context of the courses that potential clients are taking [14].

References

[1] Daqar, M. A. A., & Smoudy, A. K. (2019). The Role of Artificial Intelligence on Enhancing Customer Experience. *International Review of Management and Marketing*, 9(4), 22.

[2] Hoyer, W. D., Kroschke, M., Schmitt, B., Kraume, K., & Shankar, V. (2020). Transforming the Customer Experience through New Technologies. *Journal of Interactive Marketing*, 51, 57–71.

[3] Negroponte, N. (1995). The Digital Revolution: Reasons for Optimism. *The Futurist*, 29(6), 68.
[4] Lemke, F., Clark, M., & Wilson, H. (2006). What Makes a Great Customer Experience?
[5] Havíř, D. (2017). A Comparison of the Approaches to Customer Experience Analysis. *Economics & Business*, 31(1).
[6] Meyer, C., & Schwager, A. (2007). Measuring Customer Experience in Physical Retail Environments. *Harvard Business Review*, 85, 116–126.
[7] Kaczmarek, J., & Ryżko, D. (2009). Quantifying and Optimizing User Experience: Adapting AI Methodologies for Customer Experience Management.
[8] Pabalkar, V., & Nimbalkar, K. (2021). Impact of AI on Retail Sector in India. *European Journal of Molecular & Clinical Medicine*, 7(11), 4372–4386.
[9] Sun, Z., & Finnie, G. R. (2004). *Intelligent Techniques in E-Commerce*. Berlin: Springer.
[10] Bătăgan, L., Mărăşescu, A., & Pocovnicu, A. (2010). Consumer Rights in Digital Economy: Case Study of Romanian E-Commerce Usage. *Theoretical and Applied Economics*, 9(550), 79–96.
[11] Shahid, M. Z., & Li, G. (2019). Impact of Artificial Intelligence in Marketing: A Perspective of Marketing Professionals of Pakistan. *Global Journal of Management and Business Research*, 19(2), 1–8.
[12] Iliashenko, O., Bikkulova, Z., & Dubgorn, A. (2019). Opportunities and Challenges of Artificial Intelligence in Healthcare. In *E3S Web of Conferences* (Vol. 110, p. 02028). EDP Sciences.
[13] Hanif, M., Hafeez, S., & Riaz, A. (2010). Factors Affecting Customer Satisfaction. *International Research Journal of Finance and Economics*, 60(1), 44–52.
[14] Nichifor, E., Trifan, A., & Nechifor, E. M. (2021). Artificial Intelligence in Electronic Commerce: Basic Chatbots and the Consumer Journey. *Amfiteatru Economic*, 23(56), 87–101.
[15] Krippendorff, K. (2018). *Content Analysis: An Introduction to Its Methodology*. UK: Sage Publications, London.

Chapter 12

Application of Artificial Intelligence in Banking: A Review

Syed Hasan Jafar, Viplav Dhandhukia, and Bijay Kumar G.

Contents

12.1 Introduction

The use of artificial intelligence (AI) in banking improves efficiency of banks, saves a lot of time, contributes to cost saving and finally enhances customer experience and gives them a better value proposition in terms of better service, accessibility of various products, and related financial information.

There has been a rise in the use of AI-based tools in banking industry in recent years. AI enables banks to automate some of its activities and also helps them in better decision-making with the analysis of a lot of customer data.

With the implementation of AI, customers get a personalized service, quicker resolution, or response of their issue or requests which enhances

DOI: 10.4324/9781003206316-12

overall customer experience; a more customized offering and an overall better value proposition. For banks, it saves a lot of time and costs, increased efficiency through automation, enables them to better understand customer behavior, their needs, and preferences, with which they could come with better offerings and pursue any new opportunities; serve customers quickly and with a better personalized customer interaction.

Apart from this, primarily, banking customers require safety and security while transacting. With the increased online transactions, the fraudulent activities associated with online transactions have also increased. The banks have to bring down this risk and threats associated with online transactions. Otherwise, customers would not feel secure in carrying on their payment activities and transactions. So, AI-based tools are being used to for fraud detection. AI analyzes a large amount of data and understands spending patterns and customer behavior. If there is any odd pattern or unusual behavior like transacting from a new country, transactions at high frequency or any other unusual transactions then, AI detects and prevents such a transaction to take place. Hence, AI is also playing a key role in ensuring security which enables customers to safely carry on their financial transactions.

12.2 Literature Review

The methodology that is used to analyze the different research paper is the systematic literature review (Table 12.1). These papers show the importance of AI in banks operations.

12.3 Conclusion

We can infer that AI is advancing quicker than ever in the banking industry, which is leading to the transformation of lifestyles. Technology and creativity are filling the void left by the conventional banking system. They work in a variety of areas of banking, including core banking, customer service, organizational efficiency, and analytics, all of which contribute to better banking experience. AI aids in the speed of segregation. Initially, the pressure on experienced employees and remark AI as a threat, but after a period of adapting, they benefitted from new technology and contributed to improve job performance.

AI software will identify solutions for cyber-attack frauds and data breaches. However, the use of AI-powered technologies is still in its infancy,

Table 12.1 Systematic Literature Review

Paper	*Objective*	*Finding*
Kaur et al. (2020)	Improving functional effectiveness of banks	AI-powered mobile banking apps
	Use of robotic tools in banking processes	Use of AI-based algorithms to spot any suspicious activity and detect frauds.
	AI will help in risk management in banking including operational risk, credit risk, and market risk	Chatbots and virtual assistants to enhance customer experience by providing prompt solutions.
	AI will be helpful in finding credit worthiness of clients	Banks can incorporate bank stations which are a system of AI-based self-service terminals that provide a wide variety of banking services.
	Minimizing banking frauds and scams	
	Self-service in bank branches	
	Effective decision-making through the use of AI tools	
Indriasari, Gaol, and Matsuo (2019)	To explore the application of AI to enhance the experience of bank customers	AI can be used to understand customer behavior, patterns, and expectations. With this, banks can better cater to their customers with improvements in their product offerings and services. Also, it enables them to pursue new opportunities.
	Current challenges in application of AI in customer service	Better value proposition to customers by coming up with customized offerings after properly understanding their needs and preferences.
	Managers' perception of implementation of AI to enhance customer experience	Faster service response can be given after analyzing customer behavior and service usage pattern.
	Proposing innovative AI-based solutions that improve banking operations and enhance customer experience	AI tools enable customers to have a personalized service experience. These tools can understand customer preferences and can effectively manage their finances and goals.

(Continued)

Table 12.1 (Continued)

Paper	*Objective*	*Finding*
Kaya et al. (2019)	Application of AI in customer-focused applications to enhance customer experience	AI is being used for KYC process. AI tools evaluate the reliability of information provided by the customers and verify their identity.
	Application of AI in banking operations	AI algorithms check any suspicious activity in online transactions and prevent any fraudulent activity.
	Use of AI-based tools in financial planning and investment advisory services	The use of chatbots and virtual assistants to address any issues or requests of customers quickly without the involvement of a bank employee.
	Use of AI in wealth management for various set of clients	AI tools to give financial solutions and advisory after considering factors like customers' goals and risk-taking capability.
	Improving efficiency of banks by automating certain statutory compliances	
	Use of AI to prevent and detect any frauds in online transactions	
Ryzhkova et al. (2020)	Understanding customers' perception of implementation of AI in the banking processes	Some customers have faced difficulties due to technical failures and errors in the AI-based system.
	Areas where AI can be fully implemented after considering current AI tools and customers' readiness to adapt	AI tools may sometimes prove to be ineffective. Hence, some customers are not ready to adopt.
		Some customers reacted positively as it improves service quality in terms of getting quicker solutions and getting tasks done.
		It increases the speed of services. Automation reduces the time of routine operations and increases efficiency.
		Overall, most of them are positive with the implementation of AI in banking.

Paper	*Objective*	*Finding*
Kochhar, Purohit, and Chutani (2019)	To get the knowledge of AI application in the banking sector	AI is assuming an extremely essential part in financial area in lessening cost, alleviating hazard, distinguishing misrepresentation, and expanding consumer loyalty, however then again numerous insurances ought to be taken while executing it, as there is consistently a danger of spillage of information which could bring about an immense misfortune to the banks.
	To know the improvisation of banking sector business results with the use of AI	AI in banking—for the individuals who embrace it, AI will over the long haul give a superior encounter to clients and representatives while conveying genuine business esteem on each measurement.
	AI implementation in the banking sector	
	Analyzing the impact on Indian Banks with AI	
Donepudi (2017)	The artificial intelligence and machine learning applications and evaluation in terms of utility in different functional areas of banking	We can pertinently say that banking sector is taking over by AI and machine learning.
	Framework of institutions efficiency in terms of computational intelligence for improvisation of business	AI and ML have given the banking sector a new way of meeting their customers' demands, who are looking for smarter, convenient, and safer ways to access, save, spend, and invest their money.
Malali and Gopalakrishnan (2020)	Reason behind the Indian and financial sectors to go for AI-enabled technologies.	Fintech organizations in the nation are filling in the hole gave up by the conventional financial area, and this is occurring by utilizing the force of innovation and advancement in artificial intelligence.
	Areas where execution has to be done by AI technologies.	The clients being emphatically affected by these loaning new businesses were never viewed as workable by the conventional financial area.

(Continued)

Table 12.1 (Continued)

Paper	*Objective*	*Finding*
		AI has become the de facto technology used by all financial and technology companies to figure their platforms.
		It is assessed that inside the following decade, AI-fuelled monetary administrations will be the solitary mechanism of connection for the clients, making monetary items and loaning accessible to masses even in the distant towns of the nation, in this way making monetary incorporation a nearby reality.
Soni (2019)	Cyber security in banks	Artificial intelligence was introduced as a concept to mimic human brains.
	AI advantages and disadvantages in banking	AI technology is used for enhancement of customer communication and experience.
	AI for customer interaction	Enhancement of the efficiency in banking processes which interns develop the security and controls risk. Information about cyber-attack and the prices of solutions is provided by AI toward banks and financial institutions.
		Various issues related to fraud and data breach are identified by artificial intelligence techniques.

the way they work leads to the growth and advancement in the financial and banking sector, it is possible that the future of AI-powered technologies will result in minor losses and enhance trade with utmost.

References

Donepudi, Praveen Kumar. "Machine Learning and Artificial Intelligence in Banking." *Engineering International* 5.2 (2017): pp. 83–86.

Indriasari, Elisa, Ford Lumban Gaol, and Tokuro Matsuo. "Digital Banking Transformation: Application of Artificial Intelligence and Big Data Analytics for Leveraging Customer Experience in the Indonesia Banking Sector." *2019 8th International Congress on Advanced Applied Informatics (IIAI-AAI)* (pp. 863–868). IEEE, 2019.

Kaur, Dr, et al. "Banking 4.0: 'The Influence of Artificial Intelligence on the Banking Industry & How AI Is Changing the Face of Modern-Day Banks'." *International Journal of Management* 11.6 (2020).

Kaya, Orçun, et al. "Artificial Intelligence in Banking." *Artificial Intelligence* (2019).

Kochhar, Khyati, Harsh Purohit, and Ravisha Chutani. "The Rise of Artificial Intelligence in Banking Sector." *The 5th International Conference on Educational Research and Practice (ICERP)* (p. 127). 2019.

Malali, Anil B., and S. Gopalakrishnan. "Application of Artificial Intelligence and Its Powered Technologies in the Indian Banking and Financial Industry: An Overview." *IOSR Journal of Humanities and Social Science* 25.4 (2020): pp. 55–60.

Ryzhkova, Marina, et al. "Consumers' Perception of Artificial Intelligence in Banking Sector." *SHS Web of Conferences*. Vol. 80. EDP Sciences, 2020.

Soni, Vishal Dinesh Kumar. "Role of Artificial Intelligence in Combating Cyber Threats in Banking." *International Engineering Journal for Research & Development* 4.1 (2019): p. 7.

Chapter 13

Digital Ethics: Toward a Socially Preferable Development of AI Systems

C. Guzmán-Velásquez and J. G. Lalinde-Pulido

Contents

DOI: 10.4324/9781003206316-13

13.1 Introduction

13.1.1 Introduction

Every scientific discipline generates a discourse that enables it to face the challenges of the area. Biology, geology, and psychology, among many others, have their concepts, theories, and methods, which allow researching and understanding the phenomena they deal with. The specialization of discourses and disciplines has led to the fragmentation of knowledge and the lack of an understanding of the inter-relationships between different scientific disciplines (Morley et al., 2019). But history increasingly reminds us that a fragmented style of thinking is not the most appropriate way to approach the challenges of this century, defined by the digital revolution (Veliz, 2021).

The digital revolution has had a special impact on the way we understand certain essential concepts for life, such as time, space, and human nature (Floridi, 2007). This means that the conceptual tools that we brought are no longer enough to give an explanation and meaning to the new dynamics and digital realities (Taddeo & Floridi, 2018). Therefore, science and philosophy acquire a new task of designing concepts to encompass the new environments the information societies imply (Floridi, 2015). Therefore, aimed the so-called digital revolution, and with all the chaos that it can cause when trying to understand its effects, this article chapter to do an understanding exercise, facing a problem that is gradually affecting each of the spheres of human life and scientific disciplines: the lack of bridges between engineering, science, and ethics (De Cremer & Kasparov, 2021)

Current times are characterized by how information and communication technologies are no longer only related to our individual and social well-being; our well-being and our societies are supported by these technologies (Floridi, 2021). Thanks to them, and especially to the internet, information societies have been able to establish themselves and transcend the physical and geographic space, bringing unprecedented challenges (Floridi, 2021). It is due to the digital that new frontiers of interaction and social organization are opening, including artificial intelligence (AI) agents, which leads us to rethink our role as human beings in the digital environment we inhabit (Taddeo & Floridi, 2018; Burr et al., 2018).

As humanities and computer science begin to open spaces for conversation to explore new ways of observing, intervening, and modifying human behaviors and the contexts they inhabit, digital ethics and social data sciences emerge. At the same time, philosophy and behavioral sciences seek to advance in the task of conceptually clarifying the phenomena of digital life and eventually offer conceptual tools to intervene in behaviors and design contexts.

These theoretical and conceptual developments give a glimpse of the real impact that technologies such as AI have on the different levels of society.

Having as context a highly digitized society, in which there are more digital devices connected to the network than human beings, in which offline life merges with online life, and in which human beings interact more and more with AI systems without even realizing it (Floridi, 2015), it is necessary to clarify in greater detail the pillars of digital societies to design a better Infoesphere (Floridi, 2021). To this end, Section 13.2 identifies the role of data and algorithms in the functioning of digital societies. Data is the raw material from which products and services are developed. Algorithms are relevant in this context because they provide the technology needed to manage and interpret the data and thus achieve valuable information. And with these premises, a more comprehensive definition of what AI includes: technical approaches, social practices, and industrial infrastructures (Tsamados et al., 2020). In Section 13.3, digital ethics is addressed as a particularly relevant conceptual tool for interpreting the problems and challenges of digital societies, where the use of data and algorithms affects the well-being of the society and generates problems on a global scale (Öhman & Watson, 2019). Finally, in Section 13.4, conclusions are presented.

13.2 The Role of Data and Algorithms in Today's Digital Societies

The pillars of information societies are data and algorithms. Organizations and societies use them to make "data-based decisions," seeking better relevance, precision, and even anticipation in their actions (Provost, 2013). Thanks to this infrastructure new companies, or apps, can emerge, offering convenient solutions to their users. Uber, Google, Apple, Amazon, Rappi, and Facebook, among many others, are examples of such. Data, digital economies, AI, and some of the ethical aspects that they entail are presented next.

13.2.1 Data in Digital Economies

Data is defined as: "a symbolic representation, be it numerical, alphabetic, spatial, etc., of a quantitative or qualitative attribute or variable" (Floridi et al., 2019). Thus, data are partial descriptions of empirical facts, events, and entities. When a set of data is framed in the light of a hypothesis, theory, or question, it can be converted into information of value.

It is here where the so-called surveillance economy proposed by Shoshana Zuboff (2020) begins to work. In current digital societies, technology companies collect an abysmal amount of private data about their users, both individuals and groups, giving them unprecedented power. It is a huge concentration of power, as they intensively collect data on people, populations, and network dynamics, all over the world, without any regulation (Taddeo & Floridi, 2018; Veliz, 2021). It is necessary to clarify the collection of data, at any level, since, in most cases, these companies are not interested in analyzing the data at the individual level; but at the group level, that is individuals who share characteristics, to influence their processes decision-making (Burr & Cristianini, 2019; Floridi, 2021).

The goal is to influence user decisions, with a more serious implication in the case of "social networks" such as Facebook and Google, as they sell the insights they have generated from the vast amount of private data of their users to the highest bidder. This activity is eroding democratic structures, in addition to the autonomy of its users (Van Bavel et al., 2020) by fragmenting the shared narratives that unite societies. When talking about data, it is in the broad spectrum of the concept. That is, data generated by devices, vehicles, organizations, cities, and climate monitors, among many other sources, which, thanks to sophisticated measurement sensors, can be collected and subsequently analyzed to make more precise decisions (Öhman & Watson, 2019). Thus, the data can be used to describe phenomena, to predict possible outcomes, and to prescribe courses of action.

Having a little more clarity regarding data, let's move on to a comprehensive definition of AI, the role of algorithms in this digital infrastructure, and how organizations and societies make use of it of so-called AI to analyze and make sense of the immense amount of data (Big Data) they collect and use their findings to make decisions (Mökander & Floridi, 2021; Mökander et al., 2021).

13.2.2 Artificial Intelligence, a Look Beyond the Technical Paradigm

The last 15 years have brought incredible advances in the development of AI, but the social implications of these technologies have not been given sufficient relevance (Cowls et al., 2021; Veliz, 2021). It is a historic change, where the power and scope of these technologies are outstripping the ability to understand their impact and how they work (Burr et al., 2018). And as digital technologies and AI affect more areas of daily life, they present social, political, and environmental challenges that were previously unknown (Sunstein, 2008; Öhman & Watson, 2019; Taddeo & Floridi, 2018).

Much has been said about AI and how it could come to "dominate" and even "replace" human beings when general AI is finally achieved (Burr et al., 2018), attributing to it an omnipotent scope, and therefore, taking responsibility for its designers and creators (Watson, 2020). But the evolution of recent years has revealed how this narrative hides particular interests, especially of Silicon Valley companies (Watson, 2020; Milligan, 2018), and does not obey reality. The hype has distracted the attention of public opinion, academics, regulators, politicians, and designers who should devote their energy to propose comprehensive ways to articulate these technologies beneficially with society (Floridi, 2019; Burr, 2020).

The idea of AI will be explored from three perspectives to build a comprehensive vision (Crawford & Joler, 2018; Floridi, 2019; Mökander & Floridi, 2021):

- Technical approaches: Technical advances such as symbolic logic, expert systems, and machine learning (ML), including different techniques as deep learning or generative adversarial networks. Using the term intelligent for systems like these is a trap because they are far from being intelligent (Crawford & Joler, 2018). These are technologies that identify patterns, group, optimize, and make predictions on large amounts of data.
- Social practices: People who design these systems and decide which problems will be solved using these technologies. This power determines not only the types of technologies, including the types of algorithms and data, but also defines which populations are most benefited by these tools and which populations are excluded and even discriminated against (Crawford & Joler, 2018; Mittelstadt et al., 2016).
- Industrial infrastructures: AI is a huge infrastructure that requires a planetary computer network. The economic and environmental resources needed to maintain these gigantic infrastructures are immense. Thus, these infrastructures represent a deep concentration of power (Crawford & Joler, 2018). While a laborer in Mali mining lithium, to sell to tech companies, can earn $ 1 a day's work, Jeff Bezos, the

former CEO of Amazon, makes approximately $ 270 million a day, and they both work for the same industry (Crawford & Joler, 2018; Cowls et al., 2021).

Identifying the true costs of these technologies seems to be a pertinent and necessary concern. That is why ethical, social, and environmental implications must be addressed alongside technical issues, since these technologies have a global impact and time is not on our side (Floridi, 2021; Crawford, 2018). Therefore, digital ethics is a conceptual framework to comprehensively observe the new dilemmas that these technologies bring, such as inequity, privacy (Taddeo & Floridi, 2018; Veliz, 2021), autonomy, personal determination (Van Bavel et al., 2020), and digital sovereignty (Floridi, 2021), among others.

13.3 Digital Ethics: Its Nature and Scope

Due to the digital transformation of society, changes in relationships, ways of working, decision-making, social norms, interpersonal, and institutional trust have deep impact in individual and collective life. Thus, ethical reflection becomes increasingly relevant, especially for those professionals and designers of these technologies because they have a planetary scope (Floridi, 2021; Floridi, 2020; Mittelstadt, 2019).

Considering that the current revolution is the product of advances in digital technologies, there is a common agreement in defining current societies as information societies or digital societies (Ohman and Watson, 2019). This means that information is an asset of economic, political, and social interest in each of human actions (Floridi, 2007; Floridi et al., 2019; Floridi & Cowls, 2019; Floridi & Strait, 2020; Mittelstadt et al., 2016, Mittelstadt, 2019). Therefore, digital ethics is responsible for addressing questions such as: What characteristics does this information have? What data is stored and how is it used? How are the digital infrastructures built? How are human beings more persuasive based on the way information is presented? Who are the designers of these digital infrastructures? On what ethical and moral values are these digital infrastructures based?

Awareness about the relevance, effects, and possibilities that information and digital technologies can have in everyday life is just beginning to gain strength in our society. The gap between those who take advantage of information management and those who do not do so is widening and creating an enormous digital divide (Floridi, 2021; Mittelstadt et al., 2016). According to UNESCO, it is necessary to consider digital literacy as one of the essential skills in which it is necessary to train people for the 21st century (Öhman & Watson, 2019). Additionally, training professionals with ethical foundations are needed (Morley et al., 2021).

As Carl Ömhan, from the Oxford Internet Institute (2019), synthesizes:

> The digital revolution brings enormous opportunities to improve public and private life, and our environments, from healthcare to smart cities to global warming. Unfortunately, these opportunities come with significant ethical challenges. In particular, the extensive use of more and more data, often personal (Big Data), the increasing dependence on algorithms to analyze it to shape options and make decisions (including machine learning, artificial intelligence, and robotics), and the gradual reduction of human participation or the supervision of automatic processes, raise urgent questions about fairness, responsibility, and respect for human rights.

Thus, social preferability should be the guiding principle to achieve a solid ethical balance for any digital project with an impact on human life. So, Luciano Floridi (2019) understands digital ethics as the branch of ethics that studies and evaluates moral problems related to

- information and data, including its generation, recording, curation, processing, dissemination, exchange, and use;
- algorithms, including AI, artificial agents, ML, and robots; and
- the corresponding practices and infrastructures, including responsible innovation, programming, piracy, professional codes, and standards.

All these aimed to formulate and support morally good solutions. Therefore, a successful way to identify and evaluate socially preferable projects is to analyze them based on their results. These are successful insofar as they help reduce, mitigate, or eradicate social or environmental problems, without introducing new damages or amplifying existing ones. Cowls et al. (2021) go on to suggest that an AI-based project is socially preferable if it is designed, developed, and deployed in a way that (1) prevents, mitigates, and/or solves problems that negatively affect human life and/or the well-being of the natural world and/or (2) allows the development of socially desirable or environmentally sustainable activities and in turn (3) does not introduce new forms of damage and/or amplify existing inequities.

13.3.1 Ethics of Artificial Intelligence

Algorithms have become crucial agents for the entire digital infrastructure that digital and information societies currently inhabit. Governments, educational institutions, companies, courts of law, and hospitals depend on AI systems to make crucial decisions. And while delegation of tasks to AI systems can improve efficiency and enable new solutions, these benefits are accompanied by ethical challenges (Mökander et al., 2021). For example, these technologies can produce discriminatory results, violate individual privacy, and undermine human self-determination (Veliz, 2021; Zuboff, 2020).

Therefore, new governance and forecasting mechanisms are needed (Floridi & Strait, 2020) to help organizations to design and implement these technologies in an ethical manner, while fostering the economic and social benefits of AI (Floridi & Cowls, 2019; Floridi, 2021). As a result, organizations have launched initiatives to establish ethical principles for the adoption of socially preferable AI. Among these initiatives are the Asilomar principles for AI (2017), the Declaration for Responsible Artificial Intelligence (Van Bavel et al., 2020), the principles offered by the IEEE in its article Ethically Aligned Design (Floridi, 2020), the ethical principles offered by the European Group of Ethics in Robotics (Floridi, 2021), and the five General Ethical Principles for a code of AI of the House of Lords of the United Kingdom (Morley et al., 2021). There is a good overlap among the five sets of principles, showing that democratic values are present when human rights serve as a guide (Floridi et al., 2019).

When these sets are reviewed in the light of the four bioethical principles—beneficence, non-maleficence, autonomy, and justice, the harmony between the sets of principles is more clear (Floridi, 2020). But, according to the Digital Ethics Lab headed by Luciano Foridi, it is necessary to add a new principle that ensures the intelligibility and responsibility of the actions carried out by AI systems: explicability (Floridi & Cowls, 2019).

The following are the non-exhaustive descriptions of the principles for AI:

- Beneficence is to promote well-being, preserve dignity, and take care of the planet.
- Non-maleficence is doing no harm, particularly concerning the prevention of breaches of privacy, security, and the ability to be cautious.
- Autonomy by implementing AI, some decision-making power is ceded to these technological artifacts. Autonomy means reaching a balance between the decision-making power kept and that delegated to artificial agents.

- Justice: AI should contribute to global justice and equitable access to the benefits of digital technologies. Promoting prosperity and preserving solidarity.
- Explicability is based on two objectives: (1) intelligibility—that is, how does the system work?—and (2) responsibility, in the ethical sense—that is, who answers for the way the system works?

13.3.1.1 Ethics as a Service: Ethical Orientation for the Design of Socially Preferable AI

Ethics as a service propose the mediation of managers and software developers so that AI systems are build using socially preferable technologies and solutions. Two tensions must be constantly assessed in the search for ethically sound AI systems: the tension between too flexible and too strict. As Crawford & Joler (2018) point out, algorithmic systems are assemblages between human and non-human agents that have many non-deterministic impacts, which is why it is necessary to understand the infrastructure where the AI system operates and its relationship with laws, regulations, and standards. The other tension arises between delegated responsibility and centralized responsibility, which refers to when to request an external audit and what aspects to audit with the external auditor. External ethical audit must be a central component of any operationalization of AI ethics (Morley et al., 2021). In addition, conflicts of interest in internal audits can lead to the inability to maintain an objective opinion by the auditor, while external audits have limitations such as not having access to all relevant information about the system due to contractual or privacy terms.

In the digital and information societies where the ethics of AI serves as an infrastructure for the appropriate design of socially preferable AI systems, a methodology must be available to define and implement ethical principles in the systems, complemented with an audit system that allows the iterative and constant review of its evolution.

13.3.1.2 Ethics-Based Auditing as a Governance Mechanism for Artificial Intelligence

There is a significant gap between the theory of the ethical principles of AI and their practical application (Morley et al., 2021). Therefore, it is necessary to find ways to include the principles, defined by ethical codes and frameworks, in AI systems while constantly reviewing and following governance mechanisms that allow verifying the appropriate materialization of the principles in AI systems. Thus, ethics-based auditing is proposed as the governance mechanism that aims to close the gap between the what (ethical principles) and the how (implementation) of AI ethics, serving as a governance mechanism for verification of the implementation of ethical principles.

Ethics-based auditing does not mechanize ethics but helps to identify, visualize, and communicate the ethical and moral values that are framed in an AI system (Morley et al., 2019). The responsibility for identifying and executing the steps to ensure that AI systems are ethically sound rests on the management of the organizations that design and deploy them (Cowls et al., 2021).

In ethics-based auditing, the auditor ensures the correct questions are addressed and answered appropriately in the design of the AI systems (Morley et al., 2021). The conversation starts from principles and goes toward the directive/managerial intervention (implementation) through the product cycle. Additionally, due to the changing nature of AI systems, the audit exercise must be continuous to find out when the system may be causing damage or behaving in unexpected ways (Cowls et al., 2021).

13.4 Conclusions

Actual societies increasingly depend on digital technologies for their operation. Among the new technologies, AI systems and the data sources necessary for their operation provide great power to their designers, which can affect from individuals to entire social systems. Therefore, organizational devices such as ethics as a service are increasingly necessary for the proper materialization of ethical principles in the design and implementation of AI systems. Additionally, governance and control systems such as ethics-based auditing become pillars of the ethical infrastructure (Floridi et al., 2020) in digital societies, to design socially preferable systems and solutions. This is a field that is in the early stages of its development and that still needs time and experimentation to identify the limitations that devices and practices such as ethics as a service and ethics-based auditing may have.

These are devices that enable the governance of AI systems by providing tools for visualization and monitoring their results and seeking to transcend the black-box model. Also, once implemented internally and externally by organizations and governments, they will allow citizens and users to be kept informed of how the decisions that an AI system has reached and to submit them to new consideration. Not only seeking to alleviate human suffering and anticipate the possible harm caused by these technologies but also, and more importantly, assigning the responsibilities to the right people and organizations.

References

Burr, C., & Cristianini, N. (2019). Can Machines Read Our Minds? *Minds & Machines*, 29, 461–494. https://doi.org/10.1007/s11023-019-09497-4

Burr, C., Cristianini, N., & Ladyman, J. (2018). An Analysis of the Interaction between Intelligent Software Agents and Human Users. *Minds & Machines*, 28, 735–774. https://doi.org/10.1007/s11023-018-9479-0

Burr, C., & Morley, J. (2020). Empowerment or Engagement? Digital Health Technologies for Mental Healthcare. In: Burr C., Milano S. (eds.) *The 2019 Yearbook of the Digital Ethics Lab. Digital Ethics Lab Yearbook*. Springer, Cham. https://doi-org.ezproxy.eafit.edu.co/10.1007/978-3-030-29145-7_5

Cowls, J., Tsamados, A., Taddeo, M. et al. (2021). A Definition, Benchmark and Database of AI for Social Good Initiatives. *Nat Mach Intell*, 3, 111–115. https://doi.org/10.1038/s42256-021-00296-0

Crawford, Kate, & Joler, Vladan. (2018, September 7). Anatomy of an AI System: The Amazon Echo As An Anatomical Map of Human Labor, Data and Planetary Resources. AI Now Institute and Share Lab. Available at: https://anatomyof.ai

De Cremer, D., & Kasparov, G. (2021). The Ethical AI—Paradox: Why Better Technology Needs More and Not Less Human Responsibility. *AI Ethics*. https://doi.org/10.1007/s43681-021-00075-y

Floridi, L. (2007). A Look into the Future Impact of ICT on Our Lives. *SSRN Electronic Journal*. 23(1), 59–64.

Floridi, L. (2015). The Onlife Initiative. The Onlife Manifesto. In: *The Onlife Manifesto*. Springer, Cham. https://doi.org/10.1007/978-3-319-04093-6_2

Floridi, L. (2020). What the Near Future of Artificial Intelligence Could Be. In: Burr C., Milano S. (eds.) *The 2019 Yearbook of the Digital Ethics Lab. Digital Ethics Lab Yearbook*. Springer, Cham. https://doi-org.ezproxy.eafit.edu.co/10.1007/978-3-030-29145-7_9

Floridi, L. (2021, June 1). The European Legislation on AI: A Brief Analysis of Its Philosophical Approach. Available at SSRN: https://ssrn.com/abstract=3873273 or http://dx.doi.org/10.2139/ssrn.3873273

Floridi, L., Cath, C., & Taddeo, M. (2019). Digital Ethics: Its Nature and Scope. In: Öhman C., Watson D. (eds.) *The 2018 Yearbook of the Digital Ethics Lab. Digital Ethics Lab Yearbook*. Springer, Cham. https://doi.org/10.1007/978-3-030-17152-0_2

Floridi, L., & Cowls, J. (2019, September 20). A Unified Framework of Five Principles for AI in Society. Available at SSRN: https://ssrn.com/abstract=3831321 or http://dx.doi.org/10.2139/ssrn.3831321

Floridi, L., Cowls, J., King, T.C. et al. (2020). How to Design AI for Social Good: Seven Essential Factors. *Science and Engineering Ethics*, 26, 1771–1796. https://doi.org/10.1007/s11948-020-00213-5

Floridi, L., & Strait, A. (2020). Ethical Foresight Analysis: What It Is and Why It Is Needed? *Minds & Machines*, 30, 77–97. https://doi.org/10.1007/s11023-020-09521-y

Milligan, M. (2018, April 12). Technology and the Ethics Gap. *ABET*. Available at: www.abet.org/technology-and-the-ethics-gap/

Mittelstadt, B. (2019, November). Principles Alone Cannot Guarantee Ethical AI (May 20, 2019). *Nature Machine Intelligence*. Available at SSRN: https://ssrn.com/abstract=3391293 or http://dx.doi.org/10.2139/ssrn.3391293

Mittelstadt, B., Allo, P., Taddeo, M., Wachter, S., & Floridi, L. (2016, December). The Ethics of Algorithms: Mapping the Debate. *Big Data & Society*. doi:10.1177/2053951716679679

Mökander, J., & Floridi, L. (2021). Ethics-Based Auditing to Develop Trustworthy AI. *Minds & Machines*, 31, 323–327. https://doi.org/10.1007/s11023-021-09557-8

Mökander, J., Morley, J., Taddeo, M. et al. (2021). Ethics-Based Auditing of Automated Decision-Making Systems: Nature, Scope, and Limitations. *Science and Engineering Ethics*, 27, 44. https://doi.org/10.1007/s11948-021-00319-4

Morley, J., Elhalal, A., Garcia, F. et al. (2021). Ethics as a Service: A Pragmatic Operationalisation of AI Ethics. *Minds & Machines*, 31, 239–256. https://doi.org/10.1007/s11023-021-09563-w

Morley, J., Floridi, L., Kinsey, L., & Elhalal, A. (2019). From What to How: An Initial Review of Publicly Available AI Ethics Tools, Methods and Research to Translate Principles into Practices. *SSRN Electronic Journal*. https://doi.org/10.2139/ssrn.3830348

Morley, J., Morton, C., Karpathakis, K., Taddeo, M., & Floridi, L. (2021). Towards a Framework for Evaluating the Safety, Acceptability and Efficacy of AI Systems for Health: An Initial Synthesis. *SSRN Electronic Journal*. https://doi.org/10.2139/ssrn.3826358

Öhman, C., & Watson, D. (2019). Digital Ethics: Goals and Approach. In: Öhman C., Watson D. (eds.) *The 2018 Yearbook of the Digital Ethics Lab. Digital Ethics Lab Yearbook*. Springer, Cham. https://doi.org/10.1007/978-3-030-17152-0_1

Provost, F., & Fawcett, T. (2013). Data Science and Its Relationship to Big Data and Data-Driven Decision Making. *Big Data*, 1(1), 51–59.

Sunstein, C. (2008). Democracy and the Internet. In: Van den Hoven J., Weckert J. (eds.) *Information Technology and Moral Philosophy* (Cambridge Studies in Philosophy and Public Policy, pp. 93–110). Cambridge University Press, Cambridge. doi:10.1017/CBO9780511498725.006

Taddeo, M., & Floridi, L. (2018). How AI Can Be a Force for Good. *Science*, 361(6404), 751–752. https://doi.org/10.1126/science.aat5991

Tsamados, A., Aggarwal, N., Cowls, J., Morley, J., Roberts, H., Taddeo, M., & Floridi, L. (2020, July 28). The Ethics of Algorithms: Key Problems and Solutions. Available at SSRN: https://ssrn.com/abstract=3662302 or http://dx.doi.org/10.2139/ssrn.3662302

Van Bavel, J. J., Harris, E. A., Pärnamets, P., Rathje, S., Doell, K. C., & Tucker, J. A. (2020) Political Psychology in the Digital (Mis)Information Age: A Model of News Belief and Sharing. *Journal of Psychological Study of Social Issues*, 15, 84–113.

Veliz, C. (2021). *Privacy Is Power*. Random House, UK.

Watson, D. (2020). The Rhetoric and Reality of Anthropomorphism in Artificial Intelligence. In: Burr C., Milano S. (eds.) *The 2019 Yearbook of the Digital Ethics Lab. Digital Ethics Lab Yearbook*. Springer, Cham. https://doi-org.ezproxy.eafit.edu.co/10.1007/978-3-030-29145-7_4

Zuboff, S. (2020). The Age of Surveillance Capitalism: The Fight for a Human Future at the New Frontier of Power. *Public Affairs*.

Index

Note: Page numbers in *italics* indicate a figure and page numbers in **bold** indicate a table on the corresponding page.

B

C

D

E

N

O

P

Q

R

S

T

U

V

W

Y

Z

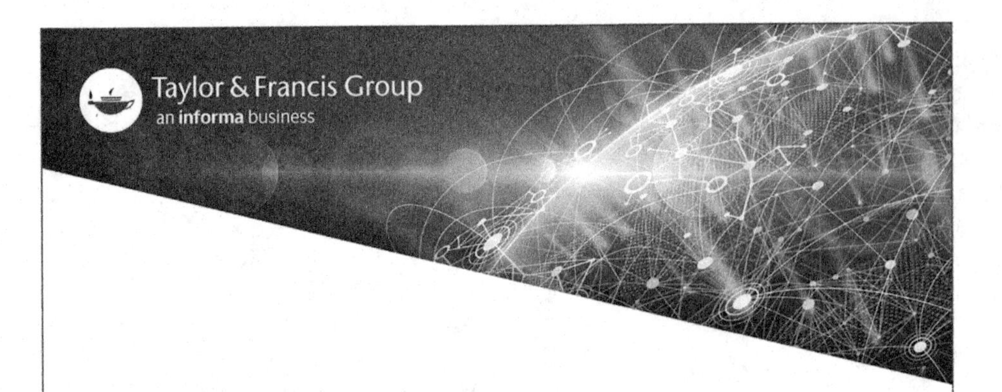

Taylor & Francis eBooks

www.taylorfrancis.com

A single destination for eBooks from Taylor & Francis with increased functionality and an improved user experience to meet the needs of our customers.

90,000+ eBooks of award-winning academic content in Humanities, Social Science, Science, Technology, Engineering, and Medical written by a global network of editors and authors.

TAYLOR & FRANCIS EBOOKS OFFERS:

A streamlined experience for our library customers

A single point of discovery for all of our eBook content

Improved search and discovery of content at both book and chapter level

REQUEST A FREE TRIAL

support@taylorfrancis.com

Printed in the United States
by Baker & Taylor Publisher Services